CARTOGRAPHY AND SITE ANALYSIS WITH MICROCOMPUTERS

A Programming Guide for
Physical Planning, Urban Design,
and Landscape Architecture

N. Brito Mutunayagam, D.E.D.P.

Ali Bahrami, M.C.R.P., M.S.

Van Nostrand Reinhold Company
New York

Copyright © 1987 by Van Nostrand Reinhold Company Inc.

Library of Congress Catalog Card Number 86-5528

ISBN 0-442-26293-0

All rights reserved. No part of this work covered by the copyright hereon may be reproduced or used in any form or by any means—graphic, electronic, or mechanical, including photocopying, recording, taping, or information storage and retrieval systems—without written permission of the publisher.

Printed in the United States of America

Designed by Joy Taylor

Van Nostrand Reinhold Company Inc.
115 Fifth Avenue
New York, New York 10003

Van Nostrand Reinhold Company Limited
Molly Millars Lane
Wokingham, Berkshire RG11 2PY, England

Van Nostrand Reinhold
480 La Trobe Street
Melbourne, Victoria 3000, Australia

Macmillan of Canada
Division of Canada Publishing Corporation
164 Commander Boulevard
Agincourt, Ontario M1S 3C7, Canada

16 15 14 13 12 11 10 9 8 7 6 5 4 3 2 1

Library of Congress Cataloging-in-Publication Data

Mutunayagam, N. Brito.
  Cartography and site analysis with microcomputers.

  Bibliography: p.
  Includes index.
  1. Building sites—Planning—Data processing.
2. Cartography—Data processing. I. Bahrami, Ali.
II. Title.
NA2540.5.M88 1986   711'.1   86-5528
ISBN 0-442-26293-0

  The authors and publisher have used their best efforts in preparing this book and the programs contained herein; this includes research, development, testing, and demonstration to verify the programs' effectiveness. The authors and publisher make no warranty of any kind, express or implied, with regard to the programs or the documentation contained in this book. The authors and publisher shall not be liable in any event for incidental or consequential damages in connection with or arising out of the delivery, performance, or use of the programs contained in this book.

Contents

Preface		v
Acknowledgments		vii
ONE	Introduction	1
TWO	The Shifting Attitude toward Computers in the Planning and Design Professions	6
THREE	Evaluating and Selecting a Microcomputer System	11
FOUR	Computer Graphics for Planning, Design, and Landscape Architecture	25
FIVE	Organization and Design of Computer Cartographic and Site Analysis Software	40
SIX	Creating and Editing a Thematic Map	55
SEVEN	Drawing Management Utilities	73
EIGHT	Simple Scanning and Opaque Overlay Utilities	93

NINE	The Multiple-Map Overlay Utility	100
TEN	A Utility for Evaluating Alternative Sites According to Specific Criteria	122
ELEVEN	The Visibility Analysis Utility	141
TWELVE	Terrain Modification and the Cut-and-Fill Utility	149
THIRTEEN	Some Concluding Thoughts	164
Appendix A	Accessing the Graphics Tablet	168
Registered Trademarks		173
Bibliography		176
Index		179

Preface

If you got beyond the cover of this book and up to this page, you must be interested in computer cartography and/or site analysis. You may even be a physical planner, urban designer, or landscape architect, who uses or has access to microcomputers. If so, this book is for you.

Most commercially available software is not necessarily developed with a planning, design, and landscape architecture audience in mind. Consequently, many packages are either entirely inapplicable or require considerable adaptation before they become fully relevant to the tasks associated with these professions. Computer cartography and site analysis are among the principal areas in which professional-quality microcomputer software is sorely lacking.

This book attempts to satisfy this need by supplying physical planners, urban designers, and landscape architects with the power and sophistication of computer graphics. It identifies some of the key graphics tasks that are involved in the creation of thematic maps and graphics data bases, and it develops and describes the software utilities needed to undertake these tasks.

The book also develops, explains, and illustrates several other computer-assisted utilities required for a variety of analyses used in determining site suitability for different purposes. All utilities are written in BASIC language, which may be readily adapted for and rendered compatible with a wide variety of microcomputer hardware configurations. As such, some programming knowledge is helpful (but not essential).

The book includes an introduction to microcomputer hardware configurations, operating systems, and applications software. It also provides guidelines for selecting an equipment-and-software configuration capable of meeting most expectations and of performing a variety of routine tasks undertaken in planning and design problem-solving settings.

Acknowledgments

Writing this book was a major undertaking that could never have been completed successfully without the support, cooperation, and assistance of a number of individuals and organizations.

The authors first wish to thank Professor Richard Austin, at whose initiation alone the writing of this book was made possible. The authors also wish to acknowledge and thank the College of Architecture of the University of Nebraska at Lincoln for having provided the environment in which the research and experimentation that culminated in the material for this book were conducted. Our special thanks go to professors Roger Massey, Gordon Scholz, Charles Deknatel, Marie Arnot, and James McGraw, to Joe Luther, and to Dean Decil Steward, without whose support, encouragement, and goodwill, this work could never have been completed.

Special acknowledgment has to be made to many students of the Department of Community and Regional Planning and the Department of Architecture for their contributions in conducting the tests and calibration of the software. The authors also wish to convey their special thanks and gratitude to Professor Lee Miller and Thomas Cheng of the Remote Sensing Laboratory of the Conservation and Survey Division of the University of Nebraska at Lincoln for their wholehearted cooperation and professional assistance in developing the presentation slides for the book. The authors wish to acknowledge and thank Mr. Leonard Campbell of Computing Services at the University of Nebraska for his guidance in using the word processor for the first draft. Thanks also to Josefa Parra, Richard W. Taylor, and Peggy Honnen for their help in compiling presentation drawings. Special thanks are due Judy R. Frederiksen who helped type, proofread, and correct the first draft manuscripts.

Finally, the authors wish to thank Carmini Brito Mutunayagam, Aneetha Mutunayagam, and Soheila Bahrami for their sacrifices, patience, and tolerance in accommodating inconveniences and late hours during the entire duration of manuscript preparation. It is unlikely that this book could have been completed without their continuous encouragement and support. The authors also wish to thank all others who directly or indirectly offered valuable advice, guidance, and encouragement during the course of this project.

ONE

Introduction

Computing devices have been used by human beings since time immemorial. The level of sophistication of these devices has kept pace with the progressive increase in diversity and complexity of problems to be solved.

COMPUTERS IN PLANNING AND DESIGN

Computers have proved useful in a variety of business and professional applications, whenever considerable data processing (with or without calculations) has to be performed. In the planning field, they have made a mark in such areas as statistical analysis and simulation/optimization models. In the design field, they have been used effectively in graphics applications and data base management. In design disciplines that depend more on intuition than on data processing and numerical computations, however, the adoption of computers has been relatively slow; during the struggle for acceptance, computers have had to contend with and overcome considerable user apprehension and skepticism.

Throughout the formative years of applied computer technology, many things hindered the adoption of computers by design offices. High capital and maintenance costs, low-quality graphics with little aesthetic appeal, and the need for programmers and training were some of the impeding factors. As a result, computers were used principally for nongraphic applications such as accounting, bookkeeping, and word processing in the larger firms.

The microcomputer has brought about a dramatic shift in the attitudes that prevailed during the 1960s and 1970s. Planners and designers are recognizing the tremendous benefits that may be reaped by harnessing microcomputer technology and bringing it to bear on their decision-making processes. This recognition is especially evident in the realm of computer graphics.

THE DOMAIN OF COMPUTER GRAPHICS IN PLANNING AND DESIGN

The most relevant recent developments in microcomputer technology applicable to physical planning and design have taken place in three areas of computer graphics: computer cartography, computer-aided drafting/design, and business graphics (statistical data representation).

Computer cartography offers many benefits in the processes of physical planning and design. Until recently, most major developments in this field involved the use of large mainframe computers. Software packages such as CMAP, SYMAP, and IMGRID pioneered the fundamental concepts that currently constitute the guidance structure for most computer-aided cartographic applications. Such applications have also been considerably enhanced by the adaptation and incorporation of computer techniques from the realms of remote sensing and photogrammetry. Many software applications programs developed in this area work with LANDSAT data in digital form.

The development of the microcomputer has led to the migration of sophisticated mapping software packages from mainframe computers to microcomputers. GRIDAPPLE, PC ATLAS, STATMAP, and PC MAP are typical packages now available for use on microcomputers.

Computer-aided drafting/design has also felt the impact of computer graphics. Computer-aided drafting enables a user to edit, alter, copy, scale, rotate, tilt, move, or eliminate different features in a drawing, while expending surprisingly little effort and time. The drafting process may occur in either two- or three-dimensional mode, usually in an interactive working environment involving the use of extensive graphic symbol libraries. Turnkey, computer-aided drafting and design systems have been developed and marketed by Calcomp, Intergraph, Computer Vision Corporation, and others, using minicomputers as hosts. More recently, computer-aided drafting software has migrated to microcomputer hosts as well. AutoCAD, AE/CADD, MICROCAD, CADKEY, and

CADPLAN are typical drafting packages available for use with microcomputers.

The third area in which computer graphics is increasingly being used is business graphics. The graphics output of such software packages as SAS/GRAPH and SPSS Graphics (which were designed for use with mainframe computers) set the trend for developing graphics packages that could generate schematic representations of statistical data. The advent of the microcomputer has made possible the development of a number of business graphics software packages for use with microcomputers. Typical examples of microcomputer-based software packages are Executive Presentation Kit, Business Graphics System, LOTUS Graphics, PFS Graph, and Microsoft CHART.

The abundance of commercially available applications software presents planners and designers with an array of graphics capabilities. In most cases, unfortunately, the source code of these packages is hidden from the user. Moreover, the documentation accompanying the software explains how to use the existing utilities but seldom describes how to perform ad hoc planning or design tasks. Most often, the software is written with an entirely different audience in mind, and the illustrations given must be adapted or modified to accommodate the needs of planners and designers. Exceptions to this pattern are few and far between, and locating them requires considerable searching. Business graphics software is perhaps the single category of applications software that requires little or no adaptation for planning and design tasks.

Several planning- and design-related tasks require graphics capabilities to be used in analytical and/or problem-solving mode. Most software packages in the mapping and drafting category require considerable adaptation (including creation of external programs and use of macros) to perform these tasks. Site analysis and site selection for a variety of purposes—areas of concern common to physical planning, urban design, landscape architecture, and architectural design—need the problem-solving capabilities of computer graphics software. Customized software is often required to undertake varied tasks of analysis and selection. In addition, some innovative procedures are needed to harness various software utilities in different sequences to make them effective in a particular situation.

OBJECTIVE OF THIS BOOK

The objective of this book is to facilitate the use of microcomputer graphics by physical planners, urban designers, and landscape architects in selected analytical applications. First, considerations for selecting an appropriate microcomputer system for graphics support in the planning and design process are reviewed. The discussion covers a range of microcomputer graphics applications software that may be adapted for use in physical planning and design.

The book then introduces a number of innovative principles for developing software utilities capable of performing many applications relevant to planning and design. Several such utilities are described and listed, and various applications are demonstrated, showing how and when they may be used for problem-solving.

Two principal procedures are emphasized in this book: creating data bases of thematic maps for applications in physical planning and urban design; and analyzing maps for site selection and display.

Thematic mapping enables users to create, store, retrieve, and update spatial information in the form of maps and legends. It also enables users to sort and scan maps and legends to locate spaces that possess certain desired attributes.

Site analysis applications that make use of thematic maps allow users to conduct analytical investigations (on the basis of specific selection criteria) necessary for the selection, modification, and development of sites and regions for specific purposes. These investigations include such things as map overlay procedures for site suitability, evaluation of alternative sites for site selection, analysis of visibility of views, cut-and-fill estimation, and optimization and statistical profiles of sites.

The routines contained in this book are written in Applesoft BASIC. They can be readily transported to other versions of BASIC (by means of appropriate conversions) or translated into other languages such as PASCAL or C. Modified versions of the routines can then be used with the appropriate hardware.

The book is intended to serve all readers from the disciplines of physical planning, urban design, and landscape architecture—irrespective of their level of computer literacy. On the one hand, a person who has no previous experience in programming can copy the routines and use the applications for performing the corresponding tasks; the only constraint in this case is that such a user must work with a microcomputer equipment configuration identical to that used by the authors (in order to avoid the inconvenience of reprogramming). On the other hand, an experienced programmer can grasp the logic of the routines and adapt them to work on a variety of hardware configurations. In either case, the demonstration examples provide the reader with procedures for using each software utility.

The authors hope that the information contained in this book will give students and professionals involved in physical planning, urban design, and landscape architecture a clearer perspective on the capabilities of relatively inexpensive microcomputer equipment in thematic mapping and site analysis applications. With the stimulus of an educated body of users, computer-assisted thematic mapping and site analysis will continue to improve their already impressive record of effectiveness, efficiency, and productivity.

TWO

The Shifting Attitude toward Computers in the Planning and Design Professions

The computer has dominated the information processing industry, virtually transforming the procedures of data collection, storage, retrieval, analyses, and information updating used in business and industry.

The use of computers in physical planning, urban design, and landscape architecture has also become an economically viable reality for professional practice. Advances in computer technology—particularly during the past two decades—have led to development of a wide variety of practical and reliable software packages, many of which are relevant to the needs of planners and designers. The concurrent advent of the personal computer enables even small planning and design firms to incorporate computer techniques into their handling of tasks and operations. Graphics capabilities of the microcomputer are particularly valuable to physical planning and design.

The current state of the art in computer hardware and software suggests a need for rapid transformation of conventional problem-solving approaches. Yet some physical planners, urban designers, and landscape architects have tended to view computer technology (and its applicability to their respective professions) with considerable apprehension and suspicion. Where computers have been adopted at all in professional practice, they have been used primarily for accounting, word processing, and bookkeeping. In other words, computers have been accepted as a useful instrument for office management but not as a tool for solving design/planning problems.

THE EARLY ANTIPATHY TOWARD COMPUTERS

Much of the early negative reaction toward computers was rooted in a fear that the new technology threatened to turn conventional professional practice into an automated and noncreative series of activities. Defensive professionals contended that design and planning were far too complex and dependent upon continual subjective and intuitive judgment to be appropriate objects of an automated or computerized process. The logic of their concern was vindicated during the 1960s and 1970s (particularly in the realm of computer graphics) by the large number of simplistic and incompetent plans and designs that were churned out by computer. These products reflected insufficient consideration of the myriad factors that must be accounted for in producing workable designs, and they displayed no improvement whatsoever over their manually developed counterparts, in either aesthetic appeal or productivity.

This situation discouraged potential users from investing the time, resources, and energy needed to acquire computers and to develop skill in the judicious selection and application of appropriate software. Experiences with computer technology of this period confirmed the notion that certain specialized skills—skills alien to those normally cultivated in an architecture, landscape architecture, urban design, or physical planning curriculum—were needed for users to operate and maintain computers productively for planning and design. Considerable data had to be gathered and processed, so proficiency in mathematics and quantitative analysis was vital; a background in analytical and solid geometry was critical for graphics applications. Finally, a completely different perspective and a fresh and open attitude toward new tools and procedures were required to facilitate a transition from the traditional drafting board to the computer keyboard.

In most instances, a prospective user had to rely on personal skill and judgment to identify, become acquainted with, and develop a working instinct for the use of appropriate software, since software normally had to be adapted by the user to handle the types of planning/design problems routinely encountered. Design and planning firms were thus obliged to invest considerable time and money to produce computer-literate individuals who possessed the computer skills needed to be effective. The downtime involved in retraining so as to use the new technology effectively was not practical for most smaller firms and professional practices, whose spare time and capital were severely limited.

The logical alternative was to hire, train, and maintain a crew of computer software consultants and programmers. Such staff members, however, were unfailingly excluded from active and direct participation in the planning and design process, confirming their status as support personnel; and the cost of maintaining them as support personnel rendered this proposition unfeasible except to very large firms and to government agencies.

Another factor in the unpopularity of the computer was that computer hardware during the 1960s and 1970s was quite expensive. Considerable space was needed to accommodate the equipment, and various sensitive components of the system could only be maintained in a controlled environment, adding additional expenses to the cost of acquisition. Once purchased, hardware required regular maintenance, as well as immediate repair whenever the machine broke down. The addition of new peripherals, together with the need to modernize obsolete equipment, necessitated specialized in-house support. The long-term and short-term financial commitments involved were thus prohibitively large for many design and planning firms.

THE EMERGING ATTITUDE OF RECENT YEARS

The advent of the microcomputer in the late 1970s and the ensuing glut of microcomputer equipment in the 1980s have completely transformed this negative scenario. A sophisticated configuration of computer equipment and software may now be acquired and assembled for just a few thousand dollars. Buyers may select from a wide range of microcomputer equipment and software, marketed at competitive prices by numerous vendors and mail-order outlets.

The presence on the market of computers at relatively low prices has made the technology available to small firms with limited budgets. Software is becoming increasingly user-friendly and now necessitates only limited knowledge of programming. Over time, the planning and design professions have become more inclined to acknowledge computers as tools capable of considerable technical sophistication and productivity; consequently, computers are gradually becoming symbols of a competitive edge among planning and design firms, rather than symbols of a soulless modernity.

DESIGN AND PLANNING TASKS SUITABLE FOR COMPUTER-AIDED ANALYSES

The tasks performed by a planner/designer can be divided into two categories: graphics applications, and nongraphics applications.

Typical graphics applications include: computer cartography, site analysis, and site selection; computer-aided design and drafting; and graphic representation of statistical data (such as with bar charts, pie charts, and data maps).

Typical nongraphic applications include (among others): data base management involving collection, organization, storage, and retrieval of data relating to projects, clients, personnel, and costs; forecasts, projections, mathematical modeling, simulation, optimization, scheduling, and other quantitative approaches to analysis and problem-solving; cost estimates, budgeting, and quantity surveying; word processing; electronic mail and data transfers; accounting and bookkeeping; and project management.

USER PREPAREDNESS AND THE PRODUCTIVE USE OF COMPUTERS

Today, microcomputer hardware may be configured and software selected to give users the capability to undertake all of the above tasks, as and when needed. But more than mere hardware and software are needed to solve complex problems. Just as essential is user preparedness.

To be successful, a user must have technical proficiency in planning or design methodology and an open attitude about using innovative tools for problem-solving.

On the technical side, the user should possess the judgment required to generate all information needed for making sound decisions. This involves having a working knowledge of the theory and methodology of problem-solving. Second, the user should be able to establish the link between the capabilities of software routines and the method used for problem-solving; this ensures development of an efficient and effective computer-aided problem-solving process, through judicious selection, application, and sequencing of appropriate software routines. Third, the user should possess the intuition required to interpret and assess the accuracy and validity of results generated by the computer.

As for attitude, the key factor is the user's sensitivity to the partnership between human being and machine. In this partnership, the user is always in command, and the machine is always subservient. The machine represents a powerful and versatile ally for creative problem-solving, but the user should not be intimidated by its power and versatility.

By combining the technical and attitudinal dimensions described above, the user may cultivate a kind of instinctive philosophy for using equipment and software judiciously, in response to the demands of the tasks at hand. This philosophy may eventually mature into a style or process of computer-aided problem-solving in planning and design.

The process so derived must be flexible and open-ended, if the creativity of the user is not to be constrained by (or restricted to) the limited operations of the hardware or software. The user should remember that every limitation encountered can be overcome through creativity and innovation. Such innovations are the ultimate measure of sophistication and originality in problem-solving, and they ensure a competitive edge in professional practice.

THREE

Evaluating and Selecting a Microcomputer System

The preceding chapter suggested that a need existed for adjustments in people's attitude toward the partnership of human beings and computers. Sufficient knowledge of the machine is a second key factor in establishing this partnership.

A microcomputer system may be visualized as an organization of interrelated component elements. In some instances, the components are integrated within a single chassis; in others, the components are separate and must be configured and assembled to constitute a system.

The anticipated capabilities of the system (chosen to match the expectations of the user) largely determine the selection and configuration of the components. Since each component of a microcomputer system performs a specific function, the user needs to become familiar with the roles and functions of each component before selecting any piece of equipment.

CLASSIFICATION OF EQUIPMENT

Microcomputers may be classified as general-purpose machines or dedicated machines. Most microcomputers tend to be designed with standard features intended to serve the needs of typical "general category" users. Features make possible computer activities ranging from performing complex "number crunching" exercises to performing animations to playing music. Some features facilitate simple word processing; others perform fairly complex sorting and matching exercises using large data bases; others enable the user to play video games for recreation; and still others perform elaborate graphics applications in two or three dimensions. Computers designed to be capable of performing all of the above functions to some degree, given the appropriate software, may be classified as general-purpose machines.

In contrast to the general-purpose microcomputers are the dedicated machines—machines that specialize in a specific application and do not perform anything else. Typical dedicated computers are those used as video game machines, automatic teller machines, or dedicated CAD systems.

Ideally, a planner/designer would be served by a custom-designed microcomputer dedicated to performing all the tasks found in the user's professional office. Realistically, however, the tasks undertaken in a planning/design office are so diverse that they cannot be gathered and classified under the dedicated category. Consequently, a general-purpose microcomputer makes a more viable, versatile, and powerful machine capable of supporting the variety of tasks that must be performed. By judicious selection of software and peripheral equipment, together with creative programming designed to facilitate tasks for which software is not available, the professional user can get maximum mileage out of the equipment that has been acquired.

This does not mean that no place exists for dedicated computers in a planning/design office. If the office's workload justifies the use of a computer for a single application, such as word processing or computer-aided drafting, the acquisition of a dedicated computer specializing in the required application is appropriate.

ELEMENTS IN A COMPUTER SYSTEM CONFIGURATION

At least 150 microcomputer models are currently being marketed in the United States. Some manufacturers market only one model; others market several, varying the combinations of features. The base unit offered by a manufacturer typically consists of a keyboard, a microprocessing unit, a video display unit, and one or more disk drives. The base unit may be composed of interconnected detachable elements, or it may appear as integrated units that consist of two or more built-in components each.

A computer configuration is usually made up of a base unit and several add-on options. A printer does not normally form part of a base unit. Printers, plotters, additional disk drives (including those for hard disks), tape drives, digitizers, light pens, video cameras, joysticks, mouses, and modems are usually classified as add-on options. Most systems may be expanded by linking one or more add-on units to the base unit. Such linkages often occur in response to demand for the special capabilities that these elements possess.

The power and adaptability of a particular configuration are determined by such features as the type of microprocessor, the memory capacity, the type and capacity of mass storage, the quality and resolution of the video display, and the configuration's capacity to perform multiple tasks and serve multiple users. The power of a configuration may be expanded by adding such items as memory, math coprocessors, special applications enhancement boards, and peripherals.

In order to configure a suitable arrangement of computer components for a professional planning/design office, a user must first understand the organization of the microcomputer and the roles and functions that each of the component elements has to fulfill.

ORGANIZATION OF A MICROCOMPUTER SYSTEM

A microcomputer system is organized as an assemblage of several functional component elements. These include input devices, microprocessor, primary memory, output devices, and mass storage devices. These elements may be linked together as shown in figure 3-1.

Communication between the user and the computer is initiated at the input device. Instructions and/or data are supplied by the user—from either the input device or mass storage—for processing by the microprocessor. The data are processed by the microprocessor in conformity with the instructions supplied, and the results are transmitted by the microprocessor to the user through the output device. The interface between the computer and the user is thus the integrated input/output

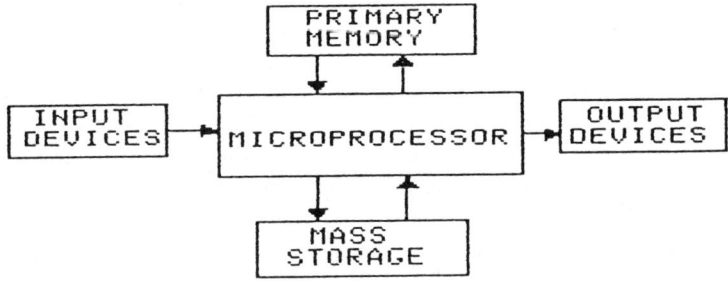

Figure 3-1. Organization of a microcomputer system.

(I/O) device, which consists of two components: the input device, and the output device.

Input Devices

The main vehicle for person-to-machine communication is the input device; this device enables the user to provide instructions and supply data to the computer. The computer processes data based on the instructions furnished to it.

Although a number of input devices may be used for person-to-machine communication, the most common one used is the keyboard, which allows entry of alphanumeric text. Instructions and/or data are supplied to the computer by entering them from the keyboard.

Digitizers (also called graphic tablets or digitizing tablets) are a very different kind of input device and are usually used for graphics applications. The tablet is a magnetic, touch-sensitive board that is usually equipped with either a digitizing pen (or stylus) or a puck. The stylus or the puck activates data entry from the tablet.

Light pens, which work in conjunction with a video monitor, constitute yet another form of input device. They operate on the same principle as the graphic tablet but use the monitor screen as the touch-sensitive surface.

One input device rapidly gaining popularity is called a mouse; it is a compact device that enables the user to select "icons" displayed on a monitor screen. (Icons are symbol displays representing different tasks.)

Output Devices

Whereas person-to-machine communication occurs by means of input devices, machine-to-person communication occurs by means of output devices. The results of data processing have to be returned to the user, and this task is performed by a variety of output devices.

The most common computer output device is the monochrome monitor. Other monitors, such as composite video monitors and RGB (Red Green Blue) monitors are available for applications that call for the use of color. Flat panel displays (which use liquid crystal or gas discharge technology) and plasma monitors may also be used as video displays in personal computers. Regular television sets can be used as output devices, but they are not suitable for use in professional offices or for any other serious applications because of their high rate of flicker and their lack of crisp images.

The display screen serves two purposes in a microcomputer configuration: to echo all data entries delivered to the microprocessor from the input devices, enabling the user to monitor the accuracy of the entries made; and to return to the user's view all results and/or errors encountered by the microprocessor during processing, based on the instructions furnished to it.

The display screen of a monitor has two modes of operation. In one mode, it serves as the equivalent of a writing pad, displaying text and numerical characters. The alphanumeric characters that represent text entries are generated on the video display by a character generator. In its other mode, the display screen serves as the canvas for graphics applications. Up to three levels of resolution are possible on a graphics monitor, depending upon the computer system it serves: low-resolution mode, medium-resolution mode, and high-resolution mode.

The monitor is principally a temporary-display-only device for output. In order to secure permanent or "hard" copies of output, the user must add a printer or plotter to the microcomputer configuration. The most common and perhaps the most inexpensive printer is the dot matrix printer. The dot matrix mechanism is made up of a matrix of impact needle points, each of which may be manipulated so that collectively the points form alphabetic, numeric, graphic, or special characters. When the dot matrix head strikes the printer's ribbon, it leaves an impression on the paper of the character demarcated by the extended needle points.

Thermal printers represent another category of printers. In operation, they follow the same principle as dot matrix printers, except that characters are formed on heat-sensitive paper.

An alternative to the dot matrix printer is the letter-quality printer, which forms characters by means of a printer head equipped with a ball element or daisy wheel bearing preformed alphabetical, numerical, and special characters. The print output of the letter-quality printer is far superior to that of the dot matrix printer, but letter-quality printers are not capable of generating true graphics output. Dot matrix printers have the advantages of being faster and more flexible for text and graphics applications.

Laser and ink jet printers are gradually becoming state of the art for hard copy printers. Both types enable users to produce high-quality text and graphics output at relatively high speeds. Ink jet printers are capable of producing hard copy in color as well.

Presentation-quality output for graphics applications may be obtained from a number of output devices besides printers. Alternative devices include pen plotters and electrostatic plotters. Plotters deliver output composed of continuous lines (rather than the dotted lines obtained from dot matrix printers). Consequently, plotted outputs are more elegant than and are preferable to printed graphics output. A variety of plotters exist, including flatbed plotters, vertical plotters, and single- and multiple-pen plotters; these come in a wide range of sizes, levels of resolution, and plotting speeds. Plotted outputs may be obtained on sheets varying in sizes from 8½ by 11 inches to sheets that are 42 inches wide.

High-quality graphics output may also be derived from camera systems, which produce high-quality transparencies for projection use; these systems are becoming increasingly popular for presentation use of computer imagery.

Linkages to Peripheral Units

Peripheral devices such as graphics tablets/digitizers, printers, plotters, modems, additional memory, and disk drives are connected to the system either by an internal system bus or by an input/output bus by interface cards.

An internal system bus is a circuit that transmits information between internal components of the microcomputer. The type of internal system buses used determines the system's flexibility to accept additional interfaces and thereby to expand its capabilities.

The input/output bus, on the other hand, is an integrated assembly of parallel electrical connections. It is used to transmit information among microcomputer and peripheral devices or among several microcomputers.

The Microprocessor

The heart of any microcomputer system is the microprocessor, which regulates the electronic data processing system and performs arithmetic and logical operations on the supplied data. It accomplishes this by breaking down a task into logical operations that can be carried out using binary numbers (zeros and ones). Every zero or one represents a "bit" of information that signals the off (zero) or on (one) position in an electronic circuit.

Assemblies of eight bits constitute a byte. One byte can represent a single alphanumeric character or one or two decimal digits. A kilobyte memory can store up to 1,024 characters.

The power of the microprocessor depends upon two factors: the word size of the microprocessor; and the clock cycle or speed of the microprocessor.

Word sizes vary from one microprocessor to another. For instance, the 6502 (which is the microprocessor installed on Apple II Plus microcomputers) is an 8-bit unit, while the Motorola 68000 (which is the microprocessor for the Apple Macintosh) and the Intel 80286 (which is the microprocessor for the IBM PC AT) are 16-bit units. As the word size gets larger, the processing task can be completed more rapidly, in fewer machine cycles.

The clock cycle is the frequency of the electronic clock that synchronizes the overall operation of the microcomputer. As the clock cycle increases, more tasks can be undertaken and completed in a shorter period of time.

User Memory

The user memory (also called primary memory) of a microcomputer is made up of semiconductors. The semiconductor memory may be subdivided into four categories: random access memory (RAM); read only memory (ROM); programmable read only memory (PROM); and erasable programmable read only memory (EPROM).

RAM, the random access memory of the microcomputer, may be visualized as the working blackboard of the computer. This memory is in an activated state from the moment the system is turned on until the system is turned off. All instructions or data delivered to the microprocessor from the input device or from mass storage reside in RAM and are available for immediate manipulation by the user.

The contents of the read only memory (ROM) are permanent, whereas those of RAM are only temporary. Programs or routines are permanently burned into ROM by the manufacturer; copies of these instructions may be recalled for use on RAM for a variety of applications, but the user may not change the instructions themselves because they are inaccessible.

The programmable read only memory (PROM) is a subclass of read only memory. PROM is usually programmed by the manufacturer; the procedure is performed only once, after which the PROM program is permanent.

Erasable programmable read only memory (EPROM) is read only memory that can be reprogrammed by the user. Reprogramming EPROM is, however, a very complicated process.

The word size, as discussed earlier, is a key determinant of computing power in a microcomputer. The larger the word size, the larger the maximum user memory that can be sustained. For instance, a typical 8-bit microprocessor can sustain a total user memory not exceeding 64K bytes, whereas a typical 16-bit microprocessor can sustain a total user memory as large as one megabyte (1,024K bytes). The volume of information that needs to be sustained on random access memory, therefore, becomes a critical element in the choice among 8-bit, 16-bit, 32-bit, and 64-bit machines.

Auxiliary Storage Devices

Auxiliary storage devices (also called mass storage devices) are peripheral equipment components of a microcomputer that enable the user to establish secondary storage for instructions and data. As described earlier, user memory is temporary. Data and instructions on user memory are lost once the electric power supply is turned off. Since they may be required again in subsequent sessions, a means to save and retrieve them at will is needed if reconstruction and reentry at every subsequent session are to be avoided. Mass storage on secondary storage devices provides such a means.

The two most common media for auxiliary storage are magnetic disks and magnetic tapes. There are two types of disks: the hard disk, and the flexible or floppy disk. The hard disk resembles a phonograph record, while the flexible disk is made of a thin film of plastic that is coated with magnetic material.

Floppy disks come in 3½-, 5¼-, and 8-inch sizes. Currently, both single-density and double-density disks are used for auxiliary storage. Information can be recorded on only one side of a single-density floppy disk, whereas it can be recorded on both sides of a double-density disk. The total capacity of a floppy disk depends upon the disk's size and density and upon the operating system.

Hard disks come in two types: removable cartridge disks, and fixed disks. Hard disks are more expensive than floppy disks, but they are considerably faster in storing and retrieving information, and they can store several megabytes of data and instructions. Microcomputers may use combinations of hard disks and floppy disks in their configuration.

Magnetic tape is also used as a medium for secondary storage. Magnetic tapes are usually used as a backup medium for data and information normally stored on a hard disk. Tapes come in spools, cartridges, or cassettes.

Hard and floppy disks must be subjected to a formatting process before they can be used as auxiliary storage media. During the formatting process, the disk is divided into tracks, and the tracks in turn are divided into sectors. Usually, the tracks on the first sector of a disk are reserved for maintaining the directory, or table of contents, of the files that will eventually be stored on the disk.

Files are saved (stored) at specific locations on the disk. The location of these files is automatically recorded in the directory. A file can be retrieved after its location address is determined from the directory.

SOFTWARE

Software may be described as a collection of instructions and/or data supplied by the user to the microprocessor to perform various tasks. Software falls into two categories: the system programs, and applications software. The most important component of system programs is the operating system.

The Operating System

The operating system manages the entire computer system and is responsible for its efficient and reliable execution of applications software. The operating system itself is a large and complex program designed to manage the hardware resources of the computer system (such as the microprocessor, auxiliary storage, and all input and output devices) and to maintain continuous communication between the user and the machine, and liaison between the machine and the applications programs.

Operating systems of various kinds are available. Among the most popular are UNIX (by Bell Laboratories), XENIX and MS DOS (by Microsoft Corporation), PC DOS (for IBM PC and IBM PC/XT microcomputers), CP/M (by Digital Research), and UCSD-P (by Softech Microsystems).

Some operating systems are more user-friendly than others, offering help menus, English-based commands, error trapping, and error recovery. Some operating systems (such as MS DOS version 2.0 and CP/M) are designed to permit a single user to perform one task at a time. Others (such as UNIX and XENIX) allow simultaneous access and task performance by multiple users. Multiuser, multitasking environments are called time-sharing systems. In a multiuser environment, the operating system takes care of regulating and channeling informational traffic, ensuring the security of each user's program and the general security of all information.

The operating system is a critical factor in a microcomputer system. Different computer systems that use the same operating system are called "compatible" with one another. This implies that applications software programmed on one compatible model may run on another, with very little modification. Conversely, an applications program created on one model of computer and compatible to one operating system will not run on an identical computer model that uses a different, incompatible operating system.

Computer Languages

Instructions furnished to a microprocessor by a user must be expressed in a language it can understand and respond to, and the only language a microprocessor is equipped to understand is machine code. Machine code is very limited in vocabulary, which makes programming in the code extremely laborious and complicated. Programs written in machine code are therefore liable to a high incidence of errors. These limitations have led to the characterization of machine code as a low-level language.

To overcome the limitations of using the low-level language, intervening communication media (called high-level languages) have been devised to enable the user to develop programs with greater ease and convenience. High-level language programming is easier for the user to understand, read, and write; it is also far more versatile than its low-level counterpart. BASIC is one of the most commonly accepted high-level languages designed for work with microcomputers. Others are FORTRAN, PASCAL, COBOL, FORTH, PROLOG, LISP, and C.

Although high-level languages are more efficient and convenient for the user/programmer, they must still be translated into machine code, in order for the microprocessor to understand the information being communicated. The software that performs such translation is called the compiler. Many operating systems come equipped with compilers for converting user instructions into machine code.

Applications Software

Applications software serves as the interface between the operating system and the user. Applications software consists of instructions written by the user or instructions assembled, packaged, and sold to users by vendors to make a microcomputer perform specific tasks.

Word processing software is a typical form of applications software; it enables the user to perform the writing and editing of documents. Computer-aided design software is another example; it enables the user to perform simple or sophisticated two-dimensional and three-dimensional drafting tasks. A third example, the spreadsheet, has gradually become one of the most popular forms of applications software; it is used for performing extremely varied computational tasks involving formulas and tabulated displays.

Data

Data constitute the raw material furnished to the microprocessor, to be processed in accordance with the instructions contained in applications software. Data are used to generate the information (output or results) on the basis of which decisions are made.

SELECTING A MICROCOMPUTER CONFIGURATION FOR A PLANNING/DESIGN OFFICE

The first step that should be taken before configuring and assembling a suitable computer system for a planning/design office is to decide whether or not to computerize. The user should first evaluate the cost-effectiveness of acquiring, installing, getting trained on, supporting, and maintaining a computer system in the given business environment.

In some instances, a justifiable alternative is to install a dedicated computer system for a single application. Typical examples are dedicated word processors and dedicated computer-aided drafting (CAD) systems. Dedicated systems are most appropriate when the business's level of departmentalization is considerable. Purchasing a general-purpose computer is justifiable when acquisition of a system for more than one application becomes cost-effective.

Once the decision has been made to computerize, the user should determine which potential tasks can and should be performed on computer. These tasks may then be classified into different categories—graphics applications, spreadsheet applications, word processing applications, data base management applications, accounting and business applications, and so on. The user may then attempt to match the different categories of applications foreseen with corresponding applications software, on a feature-by-feature basis. The best matches may occur in situations where the features of an applications software package allow the effective and reliable execution of all desired tasks. The matching process culminates in selection of applications software that provides the user with the fullest range of computer capabilities needed for peak productivity, within budgetary limits.

The "best" equipment is selected by matching equipment compatibility to the selected applications software. This implies selection of an operating system compatible with the greatest number of applications software packages chosen. If the desired applications software packages require more than one operating system, the user may have to make a choice: either to limit the selection of applications software packages to those capable of functioning on a single, given operating system, or to acquire the requisite number of operating systems to accommodate all the desired applications software packages. Cost-effectiveness should once again determine the proper choice.

The choice of operating system is also affected by the intended use of the computer within the planning/design office. If a single user or a queue of sequential users is anticipated, a single microcomputer may suffice. In this instance, it is immaterial whether a multiuser, multitasking operating system is selected or not.

On the other hand, if several users are expected to use the system simultaneously, a choice should be made between acquiring a system of several microcomputers and acquiring and establishing a time-sharing system on a single computer with several terminals and a multiuser, multitasking operating system. If the former alternative is selected, the user has the flexibility to incorporate each microcomputer into a network later on, in the event that information and data need to be shared among users at different work stations. Cost-effectiveness and compatibility with existing systems may be the key determinants in making the choice.

The volume of data to be processed and any minimum memory requirements specified for using the selected software are primary factors involved in gauging the needed specifications for the computer system. The larger the volume of data to be processed, the larger the memory. The desired processing speed determines whether the selected computer should be equipped with an 8-bit, 16-bit, 32-bit, or 64-bit microprocessor.

The selection of auxiliary storage devices is influenced by the volume of data handled, the applications software specifications, and the speed of data storage and retrieval desired. Hard disk drives supported by tape backup devices are ideal for handling large data bases. Floppy disk drives are convenient for portability of data and allow creation of inexpensive backups.

The selection of video displays is very important to the user. If the computer is to be used in the planning/design office primarily for such tasks as word processing, accounting, and design computations, a monochrome monitor is the best choice. The amber monochrome screen is considered ergonomically superior because it is less fatiguing to the eye than the green phosphor monochrome monitor.

If graphics applications represent a major priority, however, the user must take into account such considerations as need for color, level of resolution, and comparative cost differentials. RGB monitors with enhanced graphics capabilities may offer the ideal video display for color graphics applications. Composite video monitors are generally adequate and may be chosen if cost is a determining factor. If color is not a decisive factor, a monochrome monitor can prove to be very satisfactory and economical for graphics applications.

A printer is often a crucial element in a computer system. An inexpensive dot matrix printer with graphics capabilities is a minimum requirement for providing the user with a means of securing draft-quality hard copy output of program listings, program results, and graphics. A correspondence-quality printer may be needed for producing high-quality output derived from word processing and intended for inclusion in reports and correspondence.

Dot matrix and letter-quality printers may be insufficient to meet the demands of large-volume correspondence-quality or graphics output. In such cases, laser printers may provide the ideal solution, by virtue of the exceptionally crisp quality of output they deliver at very high speeds. At present, however, this option tends to be much more expensive than either of the other two.

Plotters are vital add-on devices for computer systems that are used extensively for graphics applications. Plotters were very costly in the late 1970s and early 1980s, but prices in recent years have dropped rapidly, and output quality, plot time, and resolution have dramatically improved. What may have been an unaffordable luxury in the past is now becoming cost-effective for virtually any planning/design office that uses graphics. Selection of a plotter may be made on the basis of the need for single or multiple pens, the desired level of resolution of the plotted output, and the desired plot time for large plots. Spatial considerations may help determine whether to select a flatbed plotter or a vertical plotter.

The needs of the user, together with the user's hands-on procedures and working style, become major determinants in selecting other optional add-on devices for the computer system. For instance, a numerical key pad can be a very valuable input device for users who perform extensive numerical data entry tasks. Graphics tablets using styluses, pucks, digitizing pens, light pens, or mouses are very useful additions to a computer system for which extensive graphics applications are anticipated. The particular graphics input device to choose from the range of options depends upon user preference and cultivated taste. The mouse is capable of replacing keyboard input for activating commands from menu selections, and it is considered by many to be a very user-friendly input device.

Telecommunication as a means of exchanging data and information is rapidly becoming a major dimension of computer use. Planners and designers can benefit considerably through telecommunication, by securing access to useful information networks, bulletin boards, and data bases. Users may be able to speed up communications with clients and other professionals through electronic mail. Modems, supported by appropriate telecommunication software, make telecommunication applications possible and should be included in the computer configuration.

CONCLUSION

There can be no set formula for configuring a system. The user's expectations and priorities need to be clearly defined; matching software and operating systems capable of fulfilling those expectations need to be identified, in priority order; and compatible hardware components need to be found. The user should confirm component/software compatibility before making any acquisitions.

Substantial tradeoffs may be anticipated and will inevitably be required in accommodating expectations, priorities, and costs. Typical considerations involved in the selection of equipment, software, and vendor include the following:

1. Equipment and software acquisition, maintenance, and update costs.
2. Product and software warranty policies, and product support.
3. Access to reliable repair and maintenance services.
4. Access to user groups.
5. Track records of the products.

One question that continually arises is that of timing the acquisition. "Is this the best time to buy?" "Is something better just around the corner, and should I wait?" "Are prices likely to drop?" "Am I buying equipment that will be obsolete by tomorrow?"

After careful consideration, the authors believe a fully satisfactory answer to any of the above questions cannot be made; neither to buy nor to wait is devoid of sacrifices, tradeoffs, or catches of some sort. Prices of electronic goods *are* constantly dropping, and "new and improved" hardware and software products *are* being introduced almost on a daily basis. Obsolescence is inevitable in the computer industry, owing to the tremendous competition and the rapid growth of high technology. This is part of the reason why the user must decide whether to computerize operations in the first place—before being judged a "winner" or a "loser" in the obsolescence scenario.

Careful estimation of current and future needs and expectations, including anticipating changes in such needs and expectations, is the first step to take before acquiring a computer system. The potential negative effects of obsolescence can be minimized considerably by carefully selecting the equipment and software that most efficiently accommodate these needs and changes. An open-ended system that permits adaptation and enhancements seems to be the most flexible arrangement. A closed-ended system is incapable of growing with an organization commensurately to the organization's changing needs and expectations. The extra flexibility to accommodate change may justify the extra dollars spent in acquiring an open-ended system.

FOUR

Computer Graphics for Planning, Design, and Landscape Architecture

Computer graphics can be defined as the abstract representation and display of objects, activities, or phenomena by a computer. The computer displays a visual image of a mental model of an existing or potentially real object, activity, or phenomenon on a selected output device, based on given data and instructions.

Computer graphics images do not differ from equivalent manually generated counterparts; they are duplications of the latter except that they are created in a different medium. The capability of high-speed editing and the precision and high level of resolution offered by computer-generated images are important justifications for the use of computer graphics in planning and design applications. The portability of graphic information on compact storage media makes computer graphics very convenient for users.

CLASSIFICATION OF COMPUTER GRAPHICS IMAGERY

Computer imagery may be explained by analogy to an imaginary continuum of possibilities. At one end of this continuum is the true image of an object (or activity or phenomenon), which a viewer may interpret without any ambiguity. At the other end of the continuum is an abstract image of the object, activity, or phenomenon. This abstract image may require an intervening step of clarification, to facilitate accurate interpretation by the viewer. Wolfgang Giloi classifies computer graphics into "generative graphics," "image analysis," and "cognitive graphics" to explain the realms of the field.[1]

At different times, the physical planner, urban designer, or landscape architect may focus on different points along this representational continuum. The point chosen at a given time depends upon the perceived need for computer graphics at any stage in the problem-solving process.

For instance, in the needs assessment and problem perception phase of a problem-solving exercise, the planner/designer/architect may have frequent recourse to abstract representations—such as flow diagrams and bubble diagrams—to express and communicate interpretations of the situation. Such diagrams may also be used during preliminary analysis and at the conceptual development phase. Later, at the design formulation phase and the evaluation of alternatives phase, it may be necessary to advance from such schematic abstractions to more representative abstractions—such as plans and elevations—in order to translate and represent specific interpretations and recommendations more accurately.

There is no set formula as to when images ought to be more representative and realistic, and when they ought to be schematic abstractions; this is a matter of individual style. The complexity of abstraction that can successfully be represented, however, is a direct function of the level of esoteric understanding of the viewing audience. A highly esoteric audience can visualize and negotiate quite complex levels of abstraction, whereas a nonesoteric audience may be able to visualize and negotiate only very superficial levels of abstraction or no abstraction at all.

In a typical problem-solving setting, the planner/designer negotiates varying levels of esoteric understanding and therefore needs to be able to produce image representations at varying levels of complexity. For such individuals, computer graphics offers a versatile and innovative means of image creation, representation, and communication.

CLASSIFICATION OF COMPUTER GRAPHICS APPLICATIONS

Three major categories of computer graphics applications are most relevant to the needs of physical planners, urban designers, and landscape architects: computer cartography, site analysis, and site selection; computer-aided drafting and design; and business graphics and statistical data representation.

Computer Cartography, Site Analysis, and Site Selection

Computers have gradually influenced the process of mapmaking. Manually produced maps are gradually being replaced by computer-generated images. User productivity and map-generating versatility have increased considerably. The convenience of maintaining and updating large volumes of geographic information on computer is great. The immediate accessibility of maps and other types of information for analysis and decision-making has been considerably enhanced. The computer has also made high levels of accuracy and precision in the representation of spatial data possible.

Computer cartography is essential to both planning and design primarily because both fields use spatial data in some form. The cartographic information used in typical physical planning and urban design applications may range in size and scale from site plans of small residential lots to large maps of counties, cities, regions, or states. The level of detail may vary from one application to another, depending upon the attributes that need to be considered for spatial analysis.

Two-dimensional images of spatial information—employing a variety of colors, hatching patterns, or numerical codes—may be used for attribute representation and analysis. Three-dimensional imagery may be used in such applications as terrain modeling of given topography.

The maps generated using computer cartographic software become the subjects of a number of analytical applications that would be detrimental to the selection of sites for particular purposes. Such processes as map overlaying, attribute sorting and selection, analysis of views, and earthwork optimization may be undertaken.

Computer-aided Drafting and Design

Computer-aided drafting and design represent another dimension in computer graphics applications. Most planners, designers, and landscape architects could profit from using computers to produce line drawings for plans, elevations, sections, isometric views, and perspective views. They might also benefit from using computer-aided drafting for schematic drawings, such as bubble diagrams, flow charts, and organizational diagrams.

The key difference between computer cartography and computer-aided design and drafting is that the former concentrates on the representation of attributes of areas (treated individually or collectively), while the latter focuses on precision in the representation of detail or on the schematic abstraction of an idea.

Computer-aided drafting enables the user to edit, alter, copy, scale, rotate, mirror, tilt, move, or eliminate features in a drawing in a fraction of the time required to accomplish these things manually. This considerably enhances the productivity of the planner/designer. For instance, the user may be able to alter building locations in a layout several times in a dynamic and interactive graphics mode, comparing several alternatives before choosing the most compatible arrangement. Another attribute of computer-aided drafting is the access it provides to large symbol libraries, allowing rapid creation of detailed drawings.

Several related drawings that have been prepared independently, using a common base map, can be aggregated and displayed in a variety of combinations, using the process of transparent overlaying. For instance, separate drawings may have been created for the mechanical, electrical, lighting, and heating systems of a building; composite drawings of mechanical and electrical systems or electrical and lighting systems can then be generated very rapidly by means of a simple overlay of the appropriate drawings.

Computer-aided drafting processes may be divided into interactive and noninteractive processes. The interactive process enables the user to monitor the effects of each action initiated from the input device. Any action may be modified or canceled, depending upon its observed compatibility with the end objective. The noninteractive process does not allow the user to monitor progressive actions. As a result, it is less convenient than the interactive process, since all cancellations, corrections, and modifications must be made after the entire drawing has been completed.

Two-dimensional drafting has been the conventional approach used by physical planners, urban designers, and landscape architects. Three-dimensional drafting provides the additional dimension of depth, closing the gap between abstraction and realism in the representation of objects. The third dimension adds considerably to the clarity of interpretation and inference about the organization and articulation of spaces; the viewer's imagination is not called upon to translate two-dimensional plans, elevations, and sections into mental images of three-dimensional objects, spaces, and enclosures.

Business Graphics and Statistical Data Representation

Business graphics applications are the third variety of graphics applications of special value to designers and planners. Both audiences tend to use statistical information in the decision-making process. Symbolic representations of statistical information, through such mechanisms as line graphs, bar charts, and pie charts, improve the visual support and aesthetic appeal of information displays.

SURVEY OF MAPPING SOFTWARE PACKAGES FOR MAINFRAME COMPUTERS

The earliest computers and mapping software used line printers to generate images representing geographical surfaces and terrain characteristics.

The printer head defined the smallest indivisible, graphically represented spatial unit—the cell. Different spatial attributes or data characteristics pertaining to each cell were represented graphically by different "gray levels," produced by overprinting one or more alphanumeric characters at the appropriate locations. Typical examples of such software are described below.

CMAP. Morton Scripter of the United States Census Bureau developed a package called CMAP that was capable of generating maps containing statistical representations of data, by area.[2] CMAP also produced choropleth maps, which were used in two ways: to represent geographic data coded by indivisible cell units, corresponding to the impression made by a line printer head; and to aggregate geographic entities possessing similar characteristics and to display the differences among characteristics through "gray scale" rendering (obtained by overstriking print head characters).[3]

SYMAP. SYMAP was one of the earliest computer cartography packages designed for use with a line printer. It was developed by Howard T. Fisher in the mid-1960s at the Laboratory for Computer Graphics and Spatial Analysis at Harvard University. The name SYMAP was derived from the term "SYnagraphic MAPping," which was defined as a process of "acting together graphically."[4]

SYMAP operates on the principle that any area can be graphically represented as a polygon bounded by vectors. Three types of maps can be produced using the SYMAP package: contour maps, conformant (or choropleth) maps, and proximal maps.

SYMAP is organized as an assemblage of modules or packages, each of which is to be used for a specific purpose in map production. The A–OUTLINES package is used to specify the outline of a study area. The A–CONFORMOLINES package is used to specify the outlines of each data zone (which collectively make up the total study area). The B–DATA POINTS package is used to specify the locations of all points for which data may be provided. The C–OTOLEGENDS package is used to specify relative positions, contents, numbering, and symbols in legends; this information is then placed alongside the maps generated. The D–BARRIERS package is used to specify the existence or prospective imposition of desired barriers likely to affect interpolation among data points. The E–VALUES package is used to specify the values to be assigned to data points for proximal or contour maps or to data zones for conformant maps. To change the reference order of data values in the E–VALUES package, the E1–VALUES INDEX is used. Finally, the F–MAP package is used to instruct the computer to generate maps based on the information provided by the other packages. Several additional options may be exercised in the F–MAP package.[5]

IMGRID. IMGRID is yet another cartographic package designed specifically for processing natural resource and land planning data. The IMGRID programming system was designed and developed by David F. Sinton, as a teaching tool in the Department of Landscape Architecture at the Harvard Graduate School of Design.

IMGRID is organized on the basis of a grid cell data structure. It assumes that all geographic entities can be described by aggregated square grid cells. The grid cell thus becomes the indivisible spatial unit to be used consistently as the basis for representing and displaying qualitative and thematic information. IMGRID includes features for storing, updating, retrieving, and displaying data related to the grid cell assembly.

The IMGRID programming system is partitioned into subsystems that contain discrete packages of special-purpose routines. Three such subsystems are the TOMLIN subsystem, the OMNIBUS subsystem, and the Data Management subsystem. The TOMLIN subsystem enables a user to analyze topographical and topological spatial relationships on grid cell data structures. The OMNIBUS subsystem consists of special programs for performing land analysis and plan evaluation. The Data Management subsystem is designed to extend data management capabilities and to allow program interfaces with other data systems.[6]

SYMAP and IMGRID were both developed for use on large, mainframe computers, such as the IBM 360 and the IBM 370. Nonetheless, the organization and structure of the data bases, the mapping strategies, the representation and display of spatial data, and the data processing procedures are valid for adaptation and application to microcomputers.

MAPPING SOFTWARE FOR MICROCOMPUTERS

GRIDAPPLE. GRIDAPPLE was designed and developed for microcomputers by Environmental Systems Research Institute (ESRI), and distributed by IRIS International of Landover, Maryland. The GRIDAPPLE package uses the Apple computer to generate high-quality computer maps and to perform geographic analyses. It represents a scaled-down version of ESRI's GRID software package. Natural resource data, socioeconomic and demographic data, and land use data may be input, stored, analyzed, processed, and displayed spatially by various operations of the package. Topographic and other information may be represented in either two or three dimensions, enabling the user to improve recognition and interpretation of images generated by the computer.

The GRIDAPPLE package has several capabilities, including file generation, interactive digitizing, search programs, statistical summaries, area calculation, density gaming, and multiple-overlay modeling for developing composite suitability, capability, and impact maps. Maps of any size can be directly converted and displayed in color through the process of interactive digitizing. The grid format for map information assembly is automatically derived, regardless of whether such information is entered in point, line, or polygon mode. Files can be updated interactively; grid cells may be split or merged; and large single- and multiple-variable grid files can be handled by GRIDAPPLE.

PC Atlas. PC Atlas, from Strategic Locations Planning of San Jose, California, is another microcomputer-based mapping software package. The program draws map boundaries by reading a boundary file of X-Y latitude-longitude coordinates. It then fills the boundaries with colors, shades, or hatching patterns that correspond to the statistical information for each enclosed area. The statistical information may be created by means of spreadsheet software, using geographic codes. The mapping program enables users to specify minimum and maximum values, data ranges, and zooming and magnification of selected areas.

STATMAP. STATMAP is a mapping software package developed and marketed by the Ganesa Group International of McLean, Virginia. STATMAP uses boundary files for fifty states and 3,123 counties of the United States. Its menus allow user selection of any desired state or county. Like Atlas, it reads data files created on spreadsheets external to the mapping software, using geographic codes. STATMAP, however, can also create data files internally. The software enables the user to specify maximum and minimum ranges, colors, hatching patterns, and zooming and magnification.

PCMAP. PCMAP, designed and marketed by Geo-Software of Macomb, Illinois, is another software package intended for mapping applications. PCMAP allows the user to create choropleth maps based on a file of boundary coordinates and a data file. A menu of hatch patterns offers users a variety of class patterns to select and match. PCMAP sorts data values, optimizes class limits, makes frequency distributions in the form of bar graphs, and re-creates maps stored on disk.

IMPACT OF REMOTE SENSING AND PHOTOGRAMMETRY ON COMPUTER CARTOGRAPHY

Remote sensing and photogrammetry have gradually transformed the process of computer cartography, considerably enhancing the level of precision that is possible. Digital analysis of image information gathered by remote sensing has met with considerable success in a variety of applications. Typical applications include mineral and petroleum exploration, inventorying and assessment of forest and water resources, monitoring and inventorying of croplands and rangelands, and analysis of population distribution and urban growth.

Until recently, image analysis required large, expensive systems equipped with many intricate and sophisticated devices for performing the types of tasks required for stereo compilation, drafting, plotting, digital terrain modeling, and slope generation. Low-cost digital image analysis became a possibility with the advent of the microcomputer, the appearance of affordable mass memory storage and color graphics devices, and the general availability of LANDSAT data in digital format. Typical software packages for image processing are described below.

IMPAC. IMPAC was developed by Egbert Scientific Software and marketed as an integrated system (SPECDATA) by Spectral Data Corporation of Hauppauge, New York. IMPAC (derived from the name Image Analysis Package for Microcomputers) is one of the first software packages to use microcomputer systems to perform state-of-the-art digital image analysis.[7] The software is developed for use with an Apple II microcomputer and uses LANDSAT image data as its input source. A number of analytical operations can be performed by IMPAC. These include interactive video image display, display of land classification maps, interactive evaluation of classification results, transformation of video images, and statistical and graphic representation of image information.

RIPS. RIPS, another microcomputer-based integrated package, was developed by EROS Data Center of Sioux Falls, South Dakota. Many of its features resemble those of IMPAC.

APPLEPIPS. APPLEPIPS, acronym for Apple Personal Image Processing System, was developed by the Telesys Group of Columbia, Maryland. It, too, is an image-processing software package developed for use on a microcomputer. The package incorporates image algebra routines, including averaging, differencing, multiplying, and ratio determination of pixel values.[8]

ERDAS 400. Perhaps the most sophisticated and powerful image-processing system is ERDAS 400, developed by Earth Resources Data Analysis of Atlanta, Georgia. The ERDAS 400 software is based on a Cromenco microcomputer. It includes programs for single-band and color-composite image displays, software zoom, image rectification and registration, image enhancement, and various classifications.

HOTLIPS. The Home Office Techniques for a Local Image Processing System (HOTLIPS), developed by the Nebraska Remote Sensing Center at Lincoln, Nebraska, has a similar configuration to ERDAS 400.

Other Packages. Several software packages are designed specifically for photogrammetry. The two microcomputer-based systems are the

Planning, Design, and Landscape Architecture

Apple/APP (Analytical Photogrammetry Programs), and the LMS system. Apple/APP, developed by the Department of Geography of the University of Georgia, in Athens, Georgia, facilitates such procedures as orientation of air photos/models, coordinate measurement, coordinate refinement and transformation, contour mapping, generation of digital terrain models (DTM), and statistical analyses. The LMS system, developed by Measuronics Corporation of Great Falls, Michigan, facilitates area and distance measurement, video digitizing, image overlay, and density slicing.[9]

The proliferation of software in the 1980s makes it impossible to identify every software package or development that has influenced computer cartography. Many of the basic principles characterizing the evolution of computer cartography have migrated from minicomputer-based systems to microcomputer-based systems. The grid cell, the polygon, and the line segment still act as the bases for digitizing spatial information. Conformant and contour maps still serve as the canvases on which planners, designers, and landscape architects perform their analyses. Various adaptations of the map overlay process still enable planners and designers to conduct spatial analysis and site selection. Thus microcomputer-based systems now offer the same sophistication, flexibility, and creativity that could only be obtained from large and expensive computer systems in the past.

SURVEY OF TURNKEY COMPUTER-AIDED DRAFTING AND DESIGN SYSTEMS

Computer-aided drafting and design systems are no longer unique in the computer hardware market. Decreases in hardware costs and advances in the variety and sophistication of software have created a large demand for integrated turnkey systems, motivating many vendors to manufacture them.

At the upper end of the cost spectrum are such systems as the CADAM system (by CADAM Inc.); the CalComp IGS-400, IGS-500, and System 25 (by California Computer Products); the McAuto General Drafting System (by McAuto McDonnell Douglas); the Intergraph Architectural and Engineering Application System (by Intergraph Corporation); and the Designer IV, V, and M (by Computer Vision Corporation).

Each system identified above has its own unique features but represents a typical turnkey system for computer graphics, drafting, and computer-aided design. A typical system configuration includes a central processor, one or more graphics work stations (including keyboards, joysticks, digitizing tablets, and independent picture processors), a control console, on-line and archival storage systems (including hard disk drives and tape drives), and a combination of output devices (including plotters and printers).

Each system is supported by sophisticated software for both graphics and nongraphics applications. A typical minicomputer-based CAD system configuration costs from $40,000 up. The system usually requires a protected environment, special installation, special maintenance, equipment and software updates, and esoteric training of personnel—all of which entail higher costs.

SURVEY OF MICROCOMPUTER-BASED DRAFTING SYSTEMS

The introduction of the microcomputer has encouraged a number of vendors to develop computer-aided drafting/design software, as well as dedicated turnkey systems, based on a selected host microcomputer. Some of the more popular microcomputer-based computer-aided drafting/design software packages are described below.[10]

IBM PC and PC-Compatible CAD Software

The IBM PC and PC-compatible microcomputers are the host microcomputers for a number of CAD software packages, including AutoCAD, AE/CADD, CADKEY, CADPLAN, and MICROCAD.

AutoCAD. AutoCAD, developed and marketed by AutoDesk Inc. of Mill Valley, California, is a menu-driven two-dimensional software package that may be used for producing schematic and working drawings. Drawings of any size or scale may be created interactively, using symbol libraries, text fonts, and line types created and stored earlier. The software enables the user to work with dimensioning, hatching, and text. Later versions of the package allow three-dimensional drafting.

AE/CADD. AE/CADD Master Template is an enhancement of AutoCAD, created for architectural and design applications by Archsoft Corporation. This software provides access to a comprehensive library of shapes, including doors, windows, structural components, plumbing fixtures, electrical symbols, appliances, furniture, and other drafting symbols. Specifications for each object may be recorded as each symbol is inserted, and these may be retrieved later when architectural schedules are prepared. New symbols may be created and added to the library.

Both AutoCAD and AE/CADD work with the IBM PC, IBM PC/XT, IBM PC AT, and IBM compatibles.

CADKEY. CADKEY, designed and marketed by Micro Control Systems of Vernon, Connecticut, is an interactive, menu-driven, two-dimensional and three-dimensional design and drafting software package suited for architectural and urban design applications, detailed drafting, and schematics. CADKEY possesses many of the features of AutoCAD and also includes hidden line removal in three-dimensional applications.

Its command structure can be modified to resemble many larger CAD/CAM system formats. CADKEY works with IBM PC and IBM compatibles.

CADPLAN. CADPLAN is a two-dimensional drafting system designed by Personal CAD Systems of Los Gatos, California. It is a two-dimensional drafting system intended for schematics, design development, and production drawings. It has several features similar to AutoCAD and is supported by extensive symbol libraries for architecture, HVAC, electrical, plumbing, and rendering applications. CADPLAN works with IBM PC and IBM compatibles.

MICROCAD. MICROCAD is an integrated two- and three-dimensional modeling and design system created and developed by Computer Aided Design of San Francisco, California. It permits development and editing of plans, elevations, sections, isometrics, and perspectives. It works with an integrated electronic spreadsheet that accepts data inputs from other spreadsheets. The software includes routines that allow computation and display of center of gravity and moment of inertia. This system also works with IBM PC and a number of PC compatibles.

Apple CAD Systems

The Apple microcomputer is the host for a number of CAD software packages, including CAD-1 and CAD-2 (by Chessel-Robocom Corporation of Newton, Pennsylvania); CASCADE I and II (by Cascade Graphics Development of Santa Ana, California); CADAPPLE (by T & W Systems, Inc. of Fountain Valley, California); and SPACE GRAPHICS/SPACE TABLET (by Micro Control Systems of Vernon, Connecticut).[11]

CAD-1 and CAD-2. These two packages are menu-driven two-dimensional drafting and graphics systems with color capabilities. They use symbol libraries for producing schematics, flow charts, and simple scale drawings. CAD-2 is more powerful than CAD-1, owing to its capabilities in the areas of numerical data entry and enhanced automatic dimensioning.

CASCADE I and II. CASCADE I is a computer-aided design system that performs the basic drafting tasks handled by typical drafting systems. CASCADE II is a turnkey system designed for two-dimensional graphics production; it uses a menu-driven drawing task format to create fundamental primitives. Symbols may be chosen from available symbol libraries, making quick development of drawings possible.

CADAPPLE. CADAPPLE is a turnkey drafting system that uses an Apple microcomputer with drafting software. CADAPPLE enables users to generate fully dimensioned, two-dimensional working drawings of plans, elevations, and sections of objects. The software accepts inputs through either joystick or digitizing tablet. Simultaneous graphics and text may be displayed.[12]

SPACE GRAPHICS/SPACE TABLET. SPACE GRAPHICS/SPACE TABLET is a unique, integrated hardware and software package for two- and three-dimensional drafting. The hardware includes a graphics tablet, at the top center of which is mounted a mechanical digitizing arm. The arm is pivoted on the tablet and at three other joints, each of which is equipped with a precise potentiometer for computing and translating rectangular coordinates of digitized points into equivalent polar coordinates.

The SPACE TABLET functions just as other digitizing systems do in plotting X and Y coordinates of points on a two-dimensional plane. The unique feature of the SPACE TABLET is that the user can also digitize points on the Z axis, thereby achieving true three-dimensional digitizing.

The SPACE GRAPHICS software enables the user to perform various transformations of the images of objects (such as rotation, movement, and scaling), interactive and numerical editing, and interactive assembly and unification of objects; it also allows the user to obtain perspective views, simultaneous display of three orthographic projections, and hard copies of images.

The systems described above share many comparable features; in addition, each system has its own unique features.

SURVEY OF BUSINESS GRAPHICS SOFTWARE FOR MAINFRAME COMPUTERS

The chief software packages used in conjunction with mainframe computers are SPSS Graphics (Statistical Package for the Social Sciences), by SPSS Inc., and SAS/GRAPH, by SAS Institute.

SPSS Graphics. SPSS Graphics enables SPSS and SPSS X users to generate color displays of statistical data. The procedures developed for SPSS Graphics allow the user to choose colors, patterns, titles, and legends. SPSS Graphics is integrated with SPSS X systems in such a way that the user can harness the extensive file- and data-handling capabilities of SPSS X to produce graphic output.

SAS/GRAPH. SAS/GRAPH operates on the same principles as SPSS Graphics. It is an interactive computer graphics system that may be used for producing color plots, bar charts, pie charts, block charts, graphs, plots, and three-dimensional plots. SAS/GRAPH uses the GCHART procedure to create horizontal and vertical bar charts, pie charts, and block charts. Multiple-color bars may be created, depending upon the graphics device used. Graphs and plots are derived by using the GPLOT procedure. Straight-line and smoothed-curve fittings for regression and spline techniques are also available with this procedure. The GSLIDE procedure allows insertion of a variety of type fonts and displays of varying sizes of text in an array of colors. Three-dimensional plots can be created using the G3D procedure. SAS/GRAPH requires such special types of graphics devices as display screens and plotters manufactured by TEKTRONIX, Hewlett-Packard, and CalComp.

SURVEY OF MICROCOMPUTER-BASED BUSINESS GRAPHICS

Statistical packages designed for use with microcomputers are increasingly becoming available. Typical packages that work with microcomputers are described below.

Executive Presentation Kit. Executive Presentation Kit was developed by Koala Technologies Corporation of Santa Clara, California. It may be used to produce column charts, bar charts, line charts, pie charts, text/word charts, Gantt charts, organization charts, high/low charts, and surface area charts. The program is organized with a menu containing seven selections and several subselections. It is an easy-to-use package and may be operated with a variety of input devices.

Business Graphics System. This software package is produced and marketed by Peachtree Software of Atlanta, Georgia. Business Graphics System is capable of producing most of the charts that Executive Presentation Kit can produce. The software uses a combination of menus and blank forms as the format for generating line graphs, bar charts, area and pie charts, histograms, and scatter charts from given data. The software is also capable of importing data from PeachCalc spreadsheets.

PFS: GRAPH. PFS: GRAPH is an economical business graphics software package produced by Software Publishing of Mountain View, California. It possesses simple routines for inputting data, specifying and designing the graphics display, and delivering outputs to the screen and to a variety of output devices.

LOTUS 1-2-3. The LOTUS 1-2-3 graphics package is an extremely powerful tool for generating business graphics; it works cooperatively with its spreadsheet counterpart. The software is produced by Lotus Development Corporation of Cambridge, Massachusetts. The software enables a user to create, view, and print simple bar charts, stacked bar charts, line charts, X/Y graphs, and pie graphs.

Microsoft Chart. This software package is produced and marketed by Microsoft Corporation of Bellevue, Washington. It is capable of producing a variety of column charts, bar charts, line charts, regression types, pie charts, text word charts, high/low charts, and surface area charts.

dGRAPH III. dGRAPH III is a software package designed by Fox and Geller of Elmwood Park, New Jersey. Its unique feature is its ability to import data from data bases created on Ashton Tate's dBASE III data base software. It is also capable of generating graphs using its own data entry system. The software not only creates standard bar, line, and pie graphs, it also creates pie-bar charts and multicharts.

Quite a few business graphics software packages are not surveyed in this book. Most are capable of producing combinations of business graphics diagrams, including line and bar charts, pie charts, organizational charts, and bubble diagrams. Some use data from spreadsheets, and many use data generated internally. Most of the packages work on the IBM PC and on PC compatibiles; some work on other microcomputers, as well.

CONCLUSION

A large number of graphics applications for a variety of purposes relevant to the planner/designer are currently available. Many of the original forms of applications software developed for mainframe computers or minicomputers are now being adapted for use with microcomputers. The descriptions in this chapter are far from comprehensive, and many other packages are available; before long, still more software packages are likely to enter the market.

NOTES

1. Wolfgang K. Giloi, *Interactive Computer Graphics: Data Structure, Algorithms, Languages* (Englewood Cliffs, N.J.: Prentice-Hall, 1978), pp. 3–6.
2. Morton W. Scripter, "Choropleth Maps on Small Digital Computers," *Proceedings of the Association of American Geographers* 1 (1969): 133–36.
3. Mark S. Monmonier, *Computer Assisted Cartography: Principles and Prospects* (Englewood Cliffs, N.J.: Prentice-Hall, 1982), pp. 36–37.
4. James W. Cerny, "Use of SYMAP Computer Mapping Program," *Journal of Geography* 71(3) (March 1972): 167.
5. The brief description of the packages is drawn from "Reference Manual for Synagraphic Computer Mapping: SYMAP," Version V, Draft 2, produced by the University of Nebraska at Lincoln, School of Environmental Development, Lincoln, Neb.
6. David F. Sinton, *An Introduction to I.M.G.R.I.D.: An Information Manipulation System for Grid Cell Data Structures*, Department of Landscape Architecture, Harvard Graduate School of Design, Cambridge, Mass. (June 1976), pp. I.2–I.20.
7. Dwight D. Egbert, "Low Cost Image Analysis: A Microcomputer Based System," *Computer Graphics World* 5(2) (February 1982): 58–59.

8. Roy A. Welch, Thomas R. Jordan, and E. Lynn Usery, "Microcomputers in the Mapping Sciences," *Computer Graphics World* 6(2) (February 1983): 34–36.
9. Ibid., pp. 40.
10. The description of features of AutoCAD, CADPLAN, CADKEY, and MICROCAD is compiled from "A Guide to Computer Software for Architects and Engineers," *Architectural Record* 172(12) (October 1984): 48–80.
11. The description of features of CAD-1, CAD-2, CASCADE I, and CASCADE II is compiled from "A Guide to Computer Software for Architects and Engineers," *Architectural Record* 172(12) (October 1984): 48–80.
12. The description of features is compiled from the CADAPPLE brochure by T & W Systems, Inc.

FIVE

Organization and Design of Computer Cartographic and Site Analysis Software

Developments in the field of computer graphics have occurred in four areas relevant to planning, design, and landscape architecture: computer cartography and site analysis; photogrammetry and remote sensing; computer-aided drafting and computer-aided design; and statistical data representation and business graphics. Only the first of these areas is within the scope of this book.

Computer cartography and site analysis software may be used by a planner, designer, or landscape architect in various problem-solving applications. A number of utilities must be present in this type of applications software if the user is to have the necessary range of tools and capabilities for undertaking and solving the multitude of problems that may arise. Utilities of particular value to designers and planners include:

• Generation and display of spatial information in the graphic mode—thematic maps.
• Creation and storage of graphic data bases (maps) for future retrieval of spatial information.
• Use of graphic data bases to sort and search for sites possessing selected spatial attributes for specific purposes.
• Management and updating of graphic data and information.

- Performance of weighted and nonweighted map overlays for analyses of site suitability, site selection, regional delineation, and so on.
- Comparison, analysis, evaluation, and ranking of alternative sites for development.
- Visibility analysis to locate vantage points for scenic views.
- Earthwork estimation.

The list of utilities cited above is far from exhaustive; hence, the complexity of a given problem may necessitate the use of additional specialized utilities not mentioned above. Typical examples of such utilities are those designed for terrain modeling and shortest path estimation. It is not possible to anticipate the exact range of utilities required to handle each unique problem-solving scenario that a particular company will encounter.

For purposes of manageability, only the utilities listed above will be treated at length in this book. Readers are encouraged to add other utilities to the list as and when these become essential. Considerable versatility and problem-solving capability are possible through creative combinations and judicious applications of the utilities that are mentioned above. Therefore, these utilities are used as a starting point for the discussions of computer cartography and site analysis applications that follow.

EQUIPMENT CONFIGURATION SUGGESTED FOR COMPUTER CARTOGRAPHY AND SITE ANALYSIS APPLICATIONS

Any configuration of computer equipment can operate the utilities discussed in this book, provided that it has capabilities for color graphics, digitizing tablet input, and hard copy graphics output. The equipment configuration used by the authors is not to be taken as an endorsement of the product nor of its superiority over any other system. The software developed and demonstrated in this book may be converted for use with other systems; software conversion guides for such purposes are now available in print.[1]

Any high-level language may be used for the development of the utilities mentioned above. Since most microcomputers are equipped with BASIC, the authors chose to develop the utilities in this language. An Apple II Plus with one or two disk drives was used as the host microcomputer for the development of the utilities cited above. A composite video was used because of its color graphics capabilities. (An RGB color monitor could have been used instead.) An Apple Graphics Tablet served as the input device for digitizing maps. An EPSON MX80FT with GRAF-TRAX was used as the output device. A plotter may be used instead of a printer.

THE STRATEGY FOR REPRESENTING SPATIAL DATA IN COMPUTER CARTOGRAPHY AND SITE ANALYSIS APPLICATIONS

The development of applications software for computer cartography and site analysis applications hinges on two factors: the conceptualization of graphics units that best correspond to spatial entities; and the choice of the graphics mode that best facilitates convenient data entry.

Spatial entities are usually visualized as mosaics of polygons or as matrices of grid cells. The polygon approach, corresponding to the concept utilized in SYMAP, uses longitude-latitude coordinates to encode the perimeter of the area representing the spatial entity being mapped. Characteristics of the enclosed area are assigned to the polygon, using a numerical or color code.

The rectangular grid coordinate system, characteristic of the IMGRID system, is the alternative to the polygon approach. The spatial entity being mapped is schematically represented as a two-dimensional array of rectangular grid cells. Characteristics of each area represented by a cell are assigned to the cell by means of a numerical or color code.

Both approaches have inherent advantages and disadvantages. The polygon approach is ideal for data representation of large areas with homogeneous characteristics. As areas with unique attributes become smaller in size (and as heterogeneity increases), considerable effort must be devoted to the coding of polygon boundaries for each of these areas. This becomes extremely cumbersome, and errors introduced during the coding process are difficult to discover and correct.

The grid cell is more congenial in instances where homogeneity of attributes and characteristics does not exist. The predetermined cellular organization (corresponding to a two-dimensional matrix) enables encoding of characteristics more conveniently and avoids the problem of having to contend with the boundary organization of polygons. The grid cell organization loses ground to the polygon organization in instances where large areas with homogeneous characteristics need to be graphically represented. Several adjoining cells have to be encoded with repetitious data, making data entry extremely monotonous. Data entry verification requires considerable concentration, and error detection can be extremely difficult.

Because polygon and grid mapping approaches possess inherent advantages that are complementary, a combination of the grid and polygon approaches offers the ideal strategy for representing spatial data.

The low-resolution graphics mode appears to be the most appropriate mode for cartographic and site analysis applications, for three reasons. First, the low-resolution pixel (hereinafter described as "the cell") possesses both spatial extent (area) and relative location characteristics. The cell therefore exhibits both the inherent area characteristics of the polygon and the relative location characteristics of the grid cell.

Second, low-resolution graphics mode enables the user to access and use a larger variety of colors than are available in high-resolution graphics mode. For example, low-resolution graphics mode on an Apple II Plus microcomputer offers a choice of sixteen different colors, whereas high-resolution graphics mode on the same microcomputer offers a choice of only four colors.

Third, although hatch patterns developed and displayed in high-resolution graphics mode represent an alternative to color, they consume considerably more memory than color codes and take longer to load onto a screen. Consequently, the adoption of low-resolution graphics mode conserves memory by devoting relatively little memory to data display. The bulk of memory may then be applied to perform the assortment of tasks involved in spatial analysis.

Unfortunately, low-resolution graphics mode does possess some inherent weaknesses. The low-resolution graphics cell is an indivisible unit of area. Characteristics attributed to the cell must therefore be uniformly distributed over the entire cell. Significant judgmental inaccuracies, characterized by jagged lines, tend to become apparent—particularly around the boundaries of polygons. These inaccuracies normally balance out, however, so that the actual area is represented fairly accurately in the overall reckoning. Precision (the inherent advantage of high-resolution graphics pixels) is the primary tradeoff cost of selecting the low-resolution mode.

One way to eliminate these shortcomings is by cell splitting. But in many microcomputers, the low-resolution graphics mode does not lend itself to this luxury without considerable reprogramming. An alternative to cell splitting is to magnify the source map by modifying its scale, thereby reducing the inaccuracies of jagged edges.

In spite of its drawbacks, low-resolution graphics mode is the superior mode for analytical applications. Low-resolution graphics is not used in the precision drafting mode, and precision in the representation of polygon boundaries for analytical purposes is not necessarily a critical issue. For instance, the boundary precision of polygons representing census tracts, climatic regions, and the like is not vital to the analytical processes in which they are used.

Low-resolution graphics mode is thus the preferable mode for computer cartography and site analysis applications with microcomputers because of its greater speed, its larger field of color options, its inherent capability to conserve memory, and its suitability for spatial analysis.

STRATEGY FOR USER-FRIENDLINESS IN COMPUTER CARTOGRAPHY AND SITE ANALYSIS APPLICATIONS

A key element in the popular success of any applications software is its user-friendliness. This characteristic is especially critical if the software is introduced to users who approach it with apprehension.

A psychological approach is therefore necessary in designing software utilities. In this book, instead of controlling the applications software from the keyboard, program control is transferred to the stylus/digitizing pen and the graphics tablet for data entry. This approach is based on the premise that the stylus/digitizing pen and the graphics tablet constitute a more familiar medium for drafting than a keyboard would. Map generation thus emulates the traditional manual procedures of tracing and painting.

The tablet serves as the drawing surface, corresponding to the drafting board. The graphics image appears on the video screen, which suggests physical separation of the drafting input and output media. Data entry by drafting generates the visual display and simultaneously assigns numerical codes to the spatial attributes of the spatial entities being created. But whereas the visual display is obvious, the numerical code assignment is invisible to the user. The approach suggested virtually eliminates (or at least considerably reduces) the inconvenience, time, and effort required for base map creation, data coding, and debugging of data maps.

The drafting utilities and the analytical routines of the software developed in this book are characterized by a sequence of interactive query-and-response procedures. Each utility or analytical routine is designed to be activated from the graphics tablet by the stylus. The activation of a routine initiates a series of prompts that are displayed on the screen; the user may respond to these either from the graphics tablet or from the keyboard. This approach shields an uninitiated user from complex program operations and contributes to the user's gradual acclimation to use of microcomputer technology in physical planning, urban design, and landscape architecture.

Computer Cartographic and Site Analysis Software 45

THE LAYOUT OF THE GRAPHICS TABLET TEMPLATE

The graphics tablet has been identified as the input device for computer cartography and site analysis applications. Graphics entries made on the graphics tablet are displayed on the screen. An important preliminary, therefore, is to organize a means of correlating the inputs made at the tablet with the outputs displayed on the screen. The graphics template used with the graphics tablet serves this purpose.

The template is an overlay mounted on the graphics tablet. For purposes of computer cartography and site analysis applications, the template is divided into two segments (see figure 5-1): the menu command area and the active work space.

The menu command area contains an array of menu boxes at the

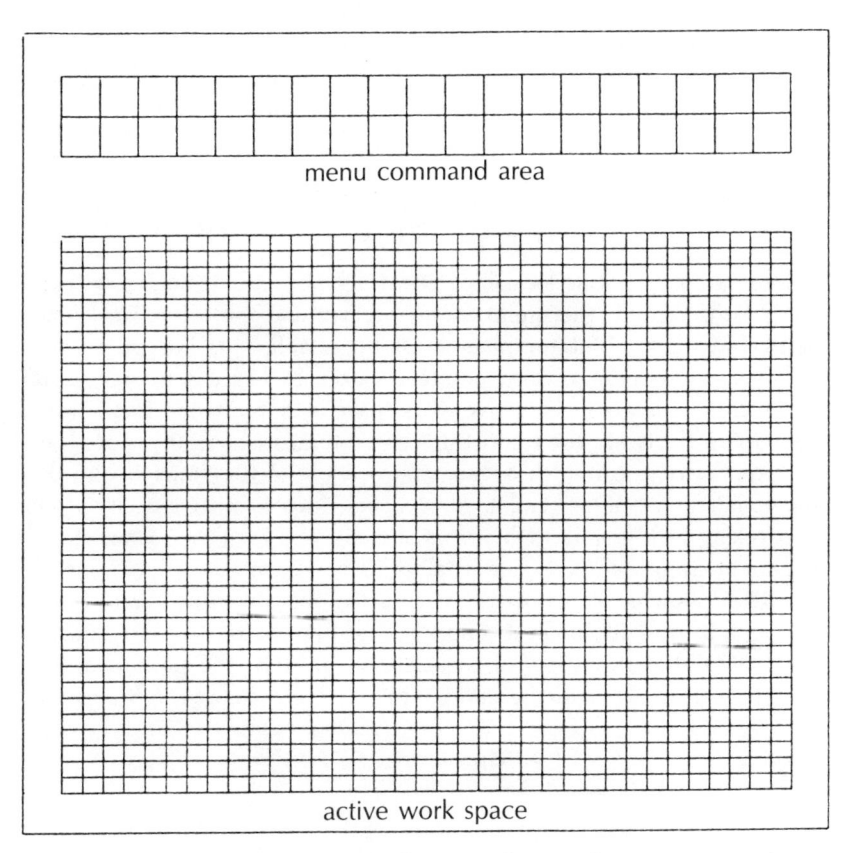

Figure 5-1. Typical design of a template.

top of the template. Each menu box may have a specific utility or analytical routine assigned to it. The utility assigned to a specific menu box may be activated by touching the tablet with the stylus at that particular menu box.

The active work space consists of the area on the tablet corresponding to the output display area on the screen. The screen may be visualized as a matrix of low-resolution graphics cells arranged in rows and columns. The active work space on the template may be organized into a corresponding matrix of cells. The bottom left-hand corner of the active work space on the template corresponds to the bottom left-hand corner of the screen. Stylus movement on the tablet is echoed by corresponding cursor movement on the screen.

The cells on the screen make discrete data representation of spatial entities possible. The position of the stylus on the template designates the location of a cell relative to the origin on the template, and corresponding to a specific location on the screen. The area of the cell displayed on the screen represents, in scale, the area of the actual spatial entity being described.

DESIGN OF THE MAIN ROUTINE FOR COMPUTER CARTOGRAPHY AND SITE ANALYSIS SOFTWARE

Before the main routine for computer cartography and site analysis software could be designed, the menu command area and the active work space of the graphics tablet template must be established. For purposes of demonstration, two rows of nineteen boxes each are set aside as the menu command area; the active work space is then divided into a matrix containing thirty-six rows and thirty-six columns (see figure 5-1).

Appendix A provides a general overview of the steps involved in programming a graphics tablet. More detailed information and instructions on working with graphics tablets may be obtained from the manufacturer's documentation accompanying a tablet.

The following program represents the main routine for computer cartography and site analysis applications. For clarity, the program may be visualized in three logical parts: the main body; the support utilities for the main body; and the cartography and site analysis subroutines.

PROGRAM LISTING

Main Body:

```
10 REM CARTOGRAPHY AND SITE ANALYSIS
   APPLICATIONS
```

Computer Cartographic and Site Analysis Software

```
20 REM              BY A. BAHRAMI
30 REM
40 DIM P%(20),Q%(21),L%(16),S(20),C$(17),
   A$(20)
50 GOSUB 900: REM INITIALIZE CONSTANTS
60 GOSUB 850: REM SETUP COLOR TABLE
70 GOSUB 800: REM INITIALIZE GRAPHIC TABLET
   VARIABLE
75 DIM D%(R4%,R4%), E%(R4%,R4%)
80 GR : REM SET THE LOW-RESOLUTION
90 GOSUB   950:   REM DRAW STATUS INDICATOR ON
   THE SCREEN
100 GOSUB 970: REM CREATING GRAPHICS DISPLAY
110 GOSUB 21000:   REM SHOW CURRENT COLOR ON
    STATUS INDICATOR
120 REM LOOP FOR EVER
130     PRINT D$;"PR#SL":PRINT "G1,S";SF;
        ",X";XF;",Y";YF;",R,B,N"
140     GOSUB 1000 : REM GET X,Y,Z FROM TABLET
160     IF YM > 1 THEN GOSUB 2000: GOTO 120
170     IF YM<=1 AND YM>=0 THEN PRINT D$;
        "PR#SL": PRINT"N": ON YM+1 GOTO 190,200
180 GOTO 120: LOOP
190 ON XM+1 GOSUB 2000,3000,4000,5000,6000,
    7000,8000,9000,10000,11000,12000,13000,
    14000,15000,16000,17000,18000,19000,20000
195 GOTO 120
200 ON XM+1 GOSUB 21000,21000,21000,21000,21000,
    21000,21000,21000,21000,21000,21000,21000,
    21000,21000,21000,21000,22000,23000,24000
210 GOTO 120
300 END
```

Support Utilities:

```
800 REM ** INITIALIZE GRAPHIC TABLET
    VARIABLES **
810 REM READ XL,YL,XH,YH,SL FROM OUTPUT OF
    TABLET INFORMATION FILE.
820 AA=3:BB=2:CC=130: REM  CONSTANTS AA, BB, CC
    CAN BE CHANGED TO SUIT THE DESIGN OF
    TEMPLATE
```

```
830 XA=XH-XL:YA=YH-YL:XF=INT
    (XA/AA):YF=INT(YA/BB):SF=INT(XA/CC)
840 RETURN
850 REM *** SETUP COLOR TABLE ***
860 C$(0)="BLACK"
861 C$(1)="MAGENTA"
862 C$(2)="DARK BLUE"
863 C$(3)="PURPLE"
864 C$(4)="DARK GREEN"
865 C$(5)="GRAY 1"
866 C$(6)="MEDIUM BLUE"
867 C$(7)="LIGHT BLUE"
868 C$(8)="BROWN"
869 C$(9)="ORANGE"
870 C$(10)="GRAY 2"
871 C$(11)="PINK"
872 C$(12)="GREEN"
873 C$(13)="YELLOW"
874 C$(14)="AQUA"
875 C$(15)="WHITE"
880 RETURN
900 REM *** INITIALIZE CONSTANTS ***
910 R4%=35:  R5%=17: S1=1: S2=1: R=0: R2=.001:
    R3=-.05: D$=CHR$(4)
920 R1%=1: DF=1: CFLAG = 1
930 EP$=CHR$(9): PL=1: XM=0
940 RETURN
950 REM *** DRAW STATUS INDICATOR ON THE
    SCREEN ***
952 REM  THIS SUB WILL INDICATE CURRENT
    SELECTED COLOR AND EDITING MODE
953 HOME
954 FOR I = 36 TO 39
956       COLOR = 0:  VLIN 0,39 AT I
958       COLOR = 15: VLIN 0,37 AT I
960 NEXT
962 COLOR=0: PLOT 38,2
963 IF PC% = 0 THEN PLOT 36,5: PLOT 37,6: PLOT
    38,6: PLOT 39,5
964 IF PC% = 1 THEN PLOT 36,6: PLOT 37,5: PLOT
    38,5: PLOT 39,6
```

```
965 FOR I=0 TO 14: COLOR=I: HLIN 37,38 AT
    I*2*8: NEXT
966 COLOR=15: HLIN 37,38 AT 39
968 RETURN
970 REM *** CREATING GRAPHICS DISPLAY ***
975 REM  THIS SUB WILL CONVERT NUMERIC DATA
    FROM MATRIX D%(R4%,R4%)
978 REM  INTO COLOR GRAPHICS AND PLOT IT ON
    THE SCREEN.
980 FOR J=R TO R4%
982    FOR I=R TO R4%
984       COLOR=D%(I,J):PLOT I,J
986    NEXT
988 NEXT
989 RETURN
1000 REM *** GET X,Y,Z FROM THE TABLET ***
1010 PRINT D$;"IN#";SL: CL=FRE(0): NORMAL
1020 POKE-16368,0: REM CLEAR THE KEYBOARD
     BUFFER
1030 INPUT X,Y,Z : REM INPUT  COORDINATES FROM
     TABLET
1040 IF Z<0 THEN GOSUB 1100:GOTO  1010:
     REM CHECK IF Esc KEY HAS BEEN PRESSED
1060 IF Z<>2 AND QQ=0 THEN 1020
1070 A=INT(X/7): B=INT(Y/4)
1080 XM=INT((A+1)/2): YM=INT((B+12)/3)
1085 IF YM<=1 AND YM>=0 THEN COLOR=15:  HLIN
     36,39 AT 1: HLIN 36,39 AT 2: HLIN 36,39
     AT 3: COLOR=0: PLOT 38,2: COLOR=DC%:
     F2%=1: RETURN
1090 F2%=0: RETURN
1100 REM *** CHECK FOR Esc KEY ***
1105 REM IF Esc KEY HAS BEEN PRESSED THEN STOP
1110 AA= PEEK(-16384): IF AA<>155 THEN RETURN
1120 PRINT D$;"PR#0: PRINT D$;"IN#0"
1130 HOME:VTAB 22 : HTAB 5
1140 PRINT "DO YOU WISH TO QUIT (Y/N)?";
1150 GOSUB 1400
1160 IF A$="Y" THEN TEXT:HOME: END
1170 PRINT: PRINT D$;"PR#";SL: PRINT "N": HOME
1180 RETURN
```

```
1200 *** CHECK TO SEE IF STYLUS IS OUTSIDE
     ACTIVE WORK SPACE ***
1210 IF  A<0  OR  A>R4% OR  B<0  OR  B>  R4%
     THEN F1%=1: RETURN
1220 F1%=0: RETURN
1300 REM *** SUBROUTINE DISK DRIVE # ***
1310 PRINT " DISK DRIVE # 1 OR 2";
1320 POKE -16368,0: GET DR$
1330 IF DR$ <>"1" AND DR$<>"2" THEN 1320
1340 RETURN
1400 REM *** SUBROUTINE RESPOND ***
1405 REM GET "Y" OR "N"
1410 POKE -16368,0: GET A$
1420 IF A$<>"Y" AND A$<>"N" THEN 1410
1420 RETURN
```

Cartography and Site Analysis Subroutine Format:

```
2000 REM *** SUBROUTINE 1 ***
2005 REM THIS SUB WILL PERFORM TASK 1
2990 RETURN
3000 REM *** SUBROUTINE 2 ***
3005 REM THIS SUB WILL PERFORM TASK 2
3990 RETURN
       •
       •
       •
24000 REM *** SUBROUTINE N ***
24005 REM THIS SUB WILL PERFORM TASK N
24990 RETURN
```

    The main body of the program extends from statement 10 to statement 300. The program continually monitors any input activity on the graphics tablet and/or keyboard. When the stylus is pressed at a chosen menu command box, it initiates the execution of the subroutine corresponding to the selected utility, or it makes the color selection assigned to that box.

    The support utilities commence at statement 800 and end at statement 1270. Each support utility performs a single specialized task. The tasks include:

1. Initializing constants.
2. Setting up the color table.
3. Initializing graphics tablet variables.
4. Drawing a status indicator on the screen.
5. Converting numerical data to color graphics.
6. Reading X, Y, and Z (sign status) values from the graphics tablet.
7. Testing the status of the Esc key.
8. Informing the user when the stylus has been pressed outside the active work space.
9. Running a user-confirmation routine based on Yes/No responses.
10. Running a verification routine to assign a disk drive.

The computer cartography and site analysis subroutines commence at statement 2000. Each subroutine performs a specific task associated with cartography and site analysis applications. A subroutine may be assigned to any convenient menu command box. As mentioned earlier, the execution of a subroutine begins when the stylus is pressed at the menu command box to which the subroutine is assigned.

DYNAMICS OF THE PROGRAM

The dynamics of the program can best be described in terms of the program's three parts (the main body, the support utilities, and the cartography and site analysis subroutines).

Dynamics of the Main Body

Lines 50, 60, and 70 call support utility subroutines 900, 850, and 800, respectively, for initializing constants, setting up the color table, and initializing the graphics tablet variables. Line 80 sets the low-resolution graphics mode. Lines 90–110 call appropriate support utility subroutines for creating graphics displays and for drawing the status indicator on the screen.

Lines 120–210 represent the main loop in the main body of the program. This loop works continuously until the Esc key is pressed. Line 130 transmits a control string to the graphics tablet (see appendix A). Line 140 secures X, Y, and Z values by calling subroutine 1000.

Lines 160–210 call the specific cartography and site analysis subroutine corresponding to the menu command box selected by the stylus. Line 190 selects any subroutine assigned to the first row of nineteen menu command boxes, while line 200 selects any subroutine assigned to the second row of nineteen menu command boxes. Table 5-1 displays the subroutine and color assignments of different menu command boxes.

TABLE 5-1.

Row #	Command Menu Box No.	Function
1	1	Cell
	2	Draw
	3	Line
	4	Polygon
	5	Protection
	6	Recode
	7	Scan
	8	Opaque Overlay
	9	Multiple Map Overlay
	10	Evaluation
	11	Contour Generation
	12	Views Analysis
	13	Earthwork Estimation
	14	Unassigned
	15	Unassigned
	16	Legend
	17	Print
	18	Clear Screen
	19	Delete
2	1	Black
	2	Magenta
	3	Dark Blue
	4	Purple
	5	Dark Green
	6	Gray 1
	7	Medium Blue
	8	Light Blue
	9	Brown
	10	Orange
	11	Gray 2
	12	Pink
	13	Green
	14	Yellow
	15	Aqua
	16	White
	17	Load
	18	Save
	19	Catalog

Computer Cartographic and Site Analysis Software

The numbering of subroutines described for various utilities in later chapters of this book is based on the organization specified in the above table. The user may modify the assignments of subroutines by rearranging the subroutine calls in lines 190 and 200.

For instance, line 190 is currently displayed as:

```
190 ON XM+1 GOSUB 2000,3000,4000,5000,6000,
    7000,8000,9000,10000,11000,12000,13000,
    14000,15000,1600,17000,18000,19000,20000
```

Subroutine 2000, which corresponds to the cell utility, is in the first position; consequently, the cell utility is assigned to the first menu command box in the first row. Subroutine 5000, which is in the fourth position in line 190, corresponds to the polygon utility; consequently, this utility is assigned to the fourth menu command box in the first row.

A user wanting to reverse these two positions need only rewrite line 190 as follows:

```
190 ON XM+1 GOSUB 5000,3000,4000,2000,6000,7000,
    8000,9000,10000,11000,12000,13000,14000,
    15000,16000,17000,18000,19000,20000
```

This action would transfer the polygon utility to the first box and the cell utility to the fourth box.

Dynamics of Support Utilities

The support utilities consist of a number of subroutines, which are described below.

Subroutine 900 initializes constants that influence the efficiency and speed of execution of the program and the flexibility to make changes to it.

Subroutine 850 sets up the color table.

Subroutine 800 reads the slot in which the graphics tablet interface card is connected. It also reads the X and Y coordinates of points on the template, which are selected during calibration (see appendix A). Line 830 defines the SCALE, XOFF, and YOFF used by the graphics tablet.

Subroutine 950 draws the status indicator along the right margin of the screen. The status indicator informs the user of the color and editing mode currently in use.

Subroutine 970 generates graphics images. The screen may be visualized as a matrix called D%(R4%,R4%), where R4% represents an equal number of rows and columns in the matrix. Every cell in the matrix has a unique location and may be assigned its own color, using a corresponding numerical code. Subroutine 970 uses the location and numerical codes to assign the desired color to the appropriate location to generate the graphics image.

Subroutine 1000 reads and interprets the values of X, Y, and Z from the graphics tablet. Line 1040 first tests Z for negative value. If the test is successful, subroutine 1100 is called. Lines 1050 and 1060 set up a flag (F2%) that will be used by the cartography and site analysis subroutines. Lines 1080 and 1085 calculate the menu coordinates XM and YM when the stylus is pressed at a menu command box.

Subroutine 1100 tests the status of the Esc key. If this key has been pressed, the user may choose either to terminate or to continue the program. This feature provides a logical escape route from the program.

Subroutine 1200 is used by cartography and site analysis subroutines. It informs the user when the stylus has been pressed outside the active work space of template.

Subroutine 1300 accepts only Y or N as a response.

Subroutine 1400 accepts only 1 or 2 as a response, to assign the disk drive that is to be accessed.

Dynamics of the Cartography and Site Analysis Subroutines

The listings, working, and demonstration applications of the cartography and site analysis subroutines are described and presented in subsequent chapters.

NOTE

1. Low-resolution graphics mode on an Apple II Plus microcomputer corresponds to text graphics mode on the IBM PC. The Apple II Plus low-resolution graphics screen has forty vertical and forty horizontal divisions. The IBM PC text mode has eighty vertical and twenty-four horizontal divisions. The Apple II Plus graphics commands translate to the following on the IBM PC:

Apple	IBM
COLOR	COLOR
PLOT	PSET,PRESET
HLIN	LINE
VLIN	LINE
SCRN	POINT

An excellent reader for converting software written in Applesoft BASIC into software for use with an IBM PC or IBM PC/XT is Richard Steck, *Apple to IBM PC Conversion Guide* (Glenview, Il.: Scott, Foresman, 1985).

S I X

Creating and Editing a Thematic Map

A thematic map is the product of a data entry process. Data entry may be performed with or without an input source map. In the latter instance, the input source map may be mounted onto the active work space of the template, as shown in figure 6-1. Digitizing is performed on the template by holding the stylus vertical to the template and touching the tablet within the active work space at the selected location.

DEALING WITH COLORS

Colors may be used in thematic mapping to represent a particular attribute or feature in graphic abstraction. As mentioned earlier, the low-resolution graphics palette on an Apple II Plus microcomputer offers the user a choice of sixteen different colors. (Other microcomputers have similar color palettes.) Each color is assigned a unique code number, by which it may be activated. For purposes of data entry, each color may be assigned to a specific box in the menu command area, as shown in table 6-1.

A selected color may be activated by the stylus, as shown in figure C-1 of the color insert. The following subroutine enables a user to select colors from the color palette:

```
21000 REM   *** SUBROUTINE COLOR ***
21010 REM THIS SUBROUTINE CHANGES STYLUS COLOR
21020 GOSUB 950: REM REDRAW STATUS INDICATOR
```

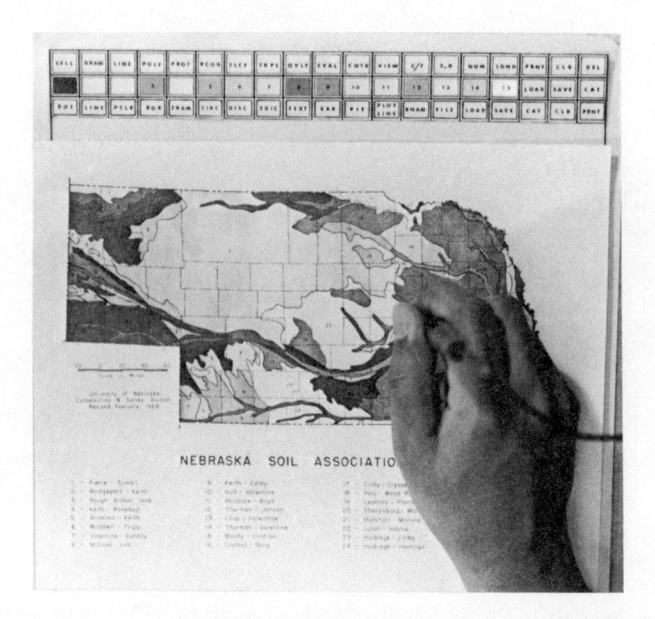

56 CARTOGRAPHY AND SITE ANALYSIS WITH MICROCOMPUTERS

Figure 6-1. Digitizing a map.

TABLE 6-1.

Color	Numerical Code
Black	0
Magenta	1
Dark Blue	2
Purple	3
Dark Green	4
Gray 1	5
Medium Blue	6
Light Blue	7
Brown	8
Orange	9
Gray 2	10
Pink	11
Green	12
Yellow	13
Aqua	14
White	15

```
21030 DC%=XM: COLOR=DC%
21040 HLIN   36,39   AT DC% * 2 + 7
21050 HLIN   36,39   AT DC% * 2 + 9
21060 HLIN   36,39   AT DC% * 2 + 8
21070 RETURN
```

Line 21020 draws the status indicator. Line 21030 activates a new color based on where the stylus is in the color menu boxes of the template. The variable *XM* is the column number of the menu box at which the stylus is positioned in the template. Therefore, if the stylus is pressed at the fifth box, *XM* will have a value of 5. In table 6-1, the numerical code 5 corresponds to Gray 1, meaning that the stylus in this case selected Gray 1 as the color for immediate use. Lines 21040–21060 highlight the color on the status indicator, as a signal to the user.

In thematic mapping, each color and its numerical code constitute a Z value that may be assigned to a cell. The color chosen for a particular cell distinguishes the attributes of that cell from the attributes of cells of other colors. By extension, the different colors signal differences in the attributes of the spatial entities that the cells represent. In doing so, they also establish a numerical score that can be interpreted in nominal, ordinal, interval, or ratio scales, as designated by the user.

The choice of color for a cell has to be made in conjunction with the determination of its location. This is because every cell has both a relative location and a specific attribute, and schematically both need to be represented together rather than independently.

BASIC DRAFTING AND DATA ENTRY UTILITIES

In order that the user may create a map, some basic drafting and data entry utilities are required. The most important of these utilities are the ones that allow graphic representation of typical spatial entities—namely, spot locations, nonlinear alignments, linear alignments, and extensive areas with homogeneous characteristics.

The utilities responsible for allowing representation of the above spatial entities are the cell utility, the draw utility, the line utility, and the polygon utility.

The Cell Utility

The cell utility enables the user to locate a spot physical feature in space at a corresponding location on a map. It also assigns a color that schematically represents the attribute of the spot location.

The following subroutine enables a user to activate the cell utility:

```
2000 REM *** SUBROUTINE CELL ***
2010 REM THIS SUBROUTINE PLOTS CELLS
2020 GOSUB    1000:IF F2%=1 THEN RETURN:REM
             RETURN IF THE STYLUS IS PRESSED AT
             ANOTHER MENU BOX
2030 GOSUB 1200: IF F1%=1 THEN 2020
2040 IF PC%=1 AND D%(A,B)<>0 THEN 2020
2050 PLOT A,B: D%(A,B)=DC%
2100 GOTO 2020
```

Line 2010 seeks out the position of the stylus. Line 2050 plots the cell in the appropriate color and saves the corresponding color code in the data matrix. Line 2040 verifies that the protection utility (discussed later in this chapter) has protected certain cell locations. If protection is in operation, data will not be plotted on those cells and will not be saved.

This subroutine necessitates a change in the main body of the main program. Line 180 must be changed to read as follows:

```
180 GOSUB 2000
```

The purpose of this modification in the main body of the main program is to create an automatic default to the cell mode.

Typical graphics output showing spot data representation created by means of the cell utility is displayed in figure 6-2.

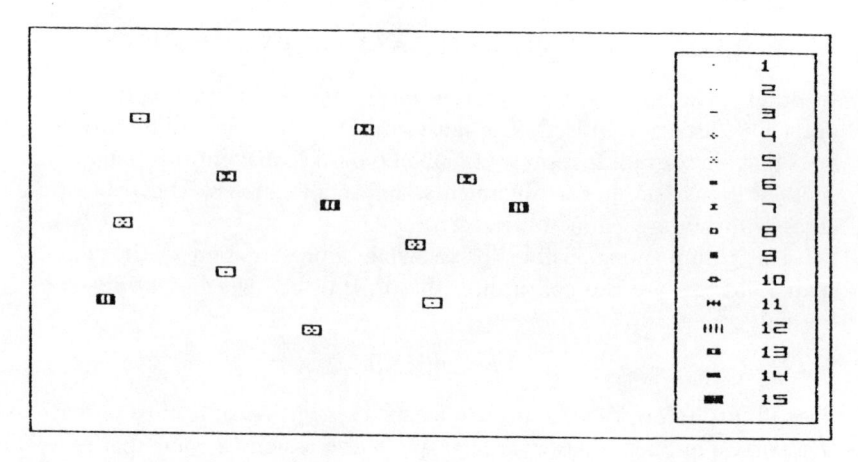

Figure 6-2. Map showing output derived from the cell utility.

The Draw Utility

The draw utility provides a convenient means of stream-digitizing nonlinear spatial entities. The cells generated by the draw utility are identical to those generated by the cell utility, except that they are sequentially linked.

The following subroutine enables a user to activate the draw utility:

```
3000 REM *** SUBROUTINE DRAW ***
3010 REM   THIS  SUBROUTINE USES STREAM MODE
     FOR DATA ENTRY FROM TABLET
3020 COLOR=15:FOR II=1 TO 3: HLIN 36,39 AT
     II:NEXT
3025 COLOR=0:PLOT 37,2:PLOT 38,1:COLOR=DC%
3030 GOSUB 1000: IF F2%=1 THEN QQ=0:RETURN
3040 QQ=1: REM SET DRAW FLAG
3050 GOSUB 1200: IF F1%=1 THEN 3030
3060 IF PC%=1 AND D%(A,B)<>0 THEN 3030
3070 PLOT A,B: D%(A,B)=DC%
3080 GOTO 3030
```

This utility resembles the cell utility, but here data entry from the graphics tablet occurs in the stream mode.

Typical graphics output showing data representation created by means of the draw utility is displayed in figure 6-3.

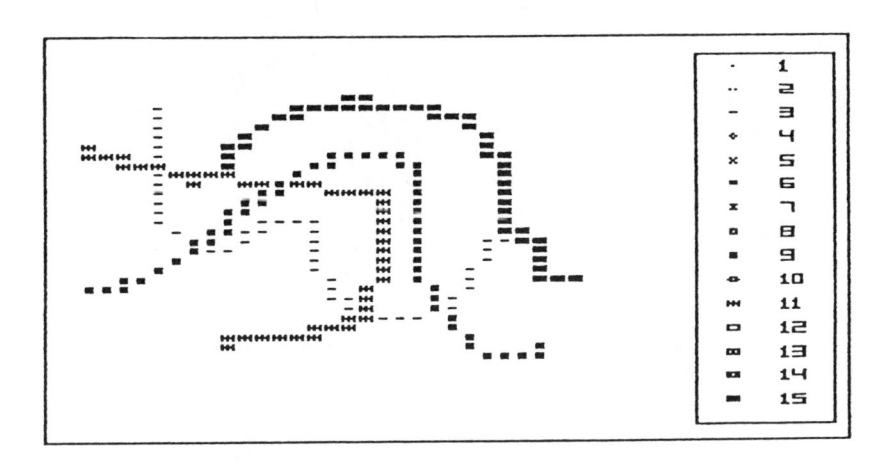

Figure 6-3. Map showing output derived from the draw utility.

The Line Utility

The line utility provides a convenient means of generating a linear array of cells to represent spatial entities that possess linear alignment characteristics. The line is defined by its starting cell and its terminating cell. A nonlinear entity may be visualized as a sequence of several linear segments, each succeeding segment of which changes direction.

Vector mathematics is employed to provide an elegant method for implementing this subroutine. A line may be drawn from any point A to any point B (see figure 6-4), once the length (L) of AB has been calculated according to the formula:

$$L = \sqrt{A^2 + B^2}$$

The calculation of the coordinates UX, UY of the unit vector U is based on the formulas:

$$UX = A/L$$

$$UY = B/L$$

This suggests that $A = L \times UX$ and $B = L \times UY$. The program to draw a line between any two points A and B functions on this principle.

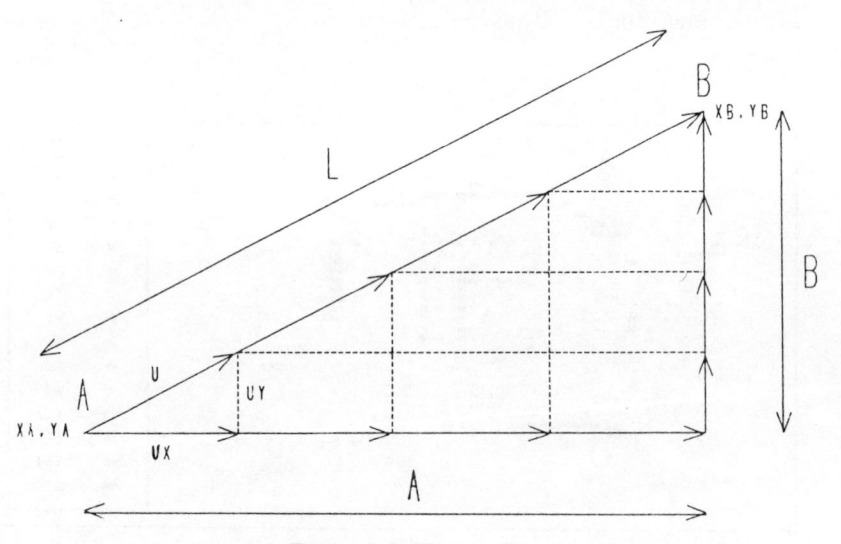

Figure 6-4. Unit vectors.

Creating and Editing a Thematic Map 61

The following subroutine enables a user to activate the line utility:

```
4000 REM *** SUBROUTINE LINE ***
4010 REM THIS SUBROUTINE DRAWS LINES USING
     VECTOR MATHEMATICS
4020 COLOR=15:FOR II=1 TO 3: HLINE 36,39 AT
     II:NEXT
4025 COLOR=0: HLIN 36,39 AT 2: COLOR = DC%
4030 GOSUB 1000: IF F2%=1 THEN RETURN
4040 GOSUB 1200: IF F1%=1 THEN 4030
4050 XB=A: YB=B
4060 IF PC%=1 AND D%(XB,YB)<>0 THEN 4080
4070 PLOT XB,YB: D%(XB,YB) = DC%
4080 GOSUB 1000: IF F2%=1 THEN RETURN
4090 GOSUB 1200: IF F1%=1 THEN 4080
4100 XE=A : YE = B
4110 IF PC% = 1 AND D%(XE,YE)<>0 THEN 4130
4120 PLOT XE,YE: D%(XE,YE) = DC%
4130 IF XE = XB AND YE = YB THEN 4080
4140 A = XE - XB: B:= YE - YB: L = SQR (A^2 +
     B^2)
4150 UX = A/L: UY = B/L
4160 FOR I=R TO L
4170       B1%=.5 + YB + I * UY: A1% =
           .5+XB+I*UX
4180       IF  PC% = 1 AND D%( A1%, B1%) <> 0
           THEN 4200
4190       PLOT A1%, B1%: D%(A1%, B1%) = DC%
4200 NEXT
4210 XB= XE: YB = YE
4220 GOTO 4080
```

Lines 4020 and 4025 indicate on the status indicator that the line utility is current. Lines 4030, 4040, and 4050 sense the coordinates of the starting point. Line 4070 plots the point. Lines 4080–4120 repeat the above steps and sense the ending point. The loop of lines 4160–4200 performs the iteration required to draw the line. Line 4210 updates the ending point to its new status as the beginning point for the next line. Line 4220 repeats the above process.

Typical graphics output showing data representation created by means of the line utility is displayed in figure 6-5.

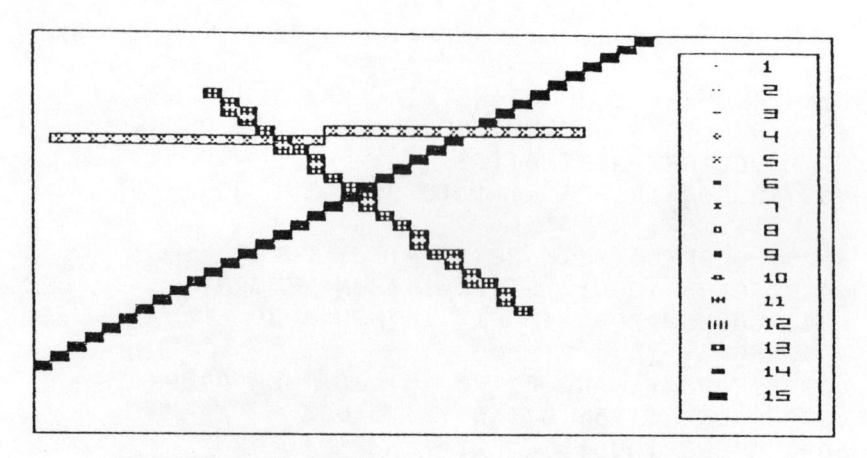

62 CARTOGRAPHY AND SITE ANALYSIS WITH MICROCOMPUTERS

Figure 6-5. Map showing output derived from the line utility.

The Polygon Utility

The polygon represents regular- or irregular-shaped spatial entities that have homogeneous characteristics. The polygon utility defines the boundary of the spatial entity, commencing from and terminating at a starting point along that boundary. It also defines the homogeneous characteristics of the enclosed area by filling in all the cells within its boundary in a selected color.

The following algorithm must be implemented to fulfill the polygon utility:

1. The boundary of the polygon is drawn on the basis of user-supplied vertices.
2. The scale at which this polygon was originally drawn is reduced by a specified increment. This scale becomes the new scale for generating an inner polygon.
3. The inner polygon is drawn.
4. Steps 2 and 3 are repeated until the polygon is fully painted.

The scale of the polygon is changed by calculating the coordinates CX,CY of the center of the polygon, using the following formulas:

$$SX = \sum_{i=1}^{n} [XB(i) + XE(i)]$$

$$SY = \sum_{i=1}^{n} [YB(i) + YE(i)]$$

In the above equations, $XB(i), YB(i)$ represent the coordinates of the beginning vertex of the line, and $XE(i), YE(i)$ represent the ending vertex of the line. SX represents the sum of X coordinates of all points, and SY represents the sum of Y coordinates of all points.

$$CX = SX/2N$$

$$CY = SY/2N$$

N represents the number of vertices.

Once the center of the polygon has been calculated (see figure 6-6), the coordinates X,Y of any point on the polygon may be computed by means of the following formulas:

$$X = [(XA - CX) SC] + CX$$

$$Y = [(YA - CY) SC] + CY$$

In the above equations, SC represents the scale factor.

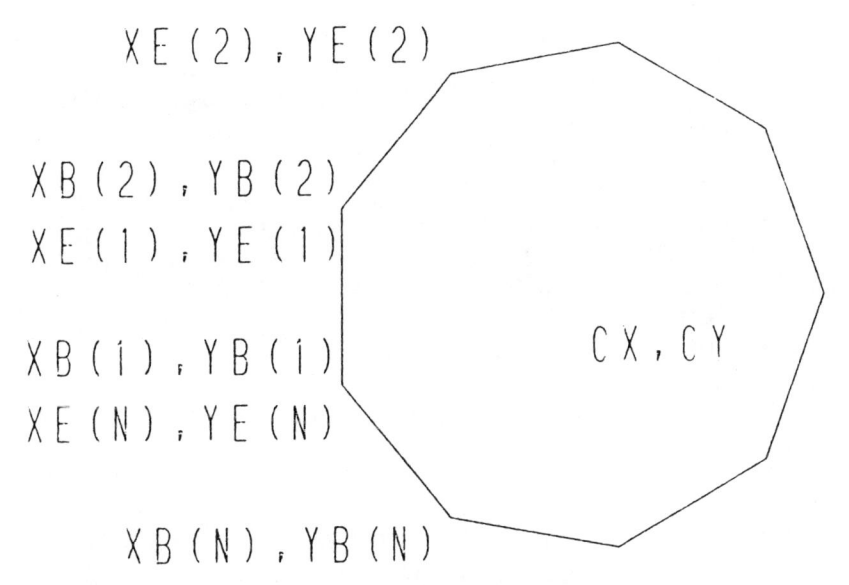

Figure 6-6. Polygon coordinates.

The following subroutine enables a user to activate the polygon utility:

```
5000 REM *** SUBROUTINE POLYGON ***
5010 REM THIS SUBROUTINE PAINTS THE POLYGON
5020 COLOR=15:FOR II=1 TO 3: HLIN 36,39 AT
     II:NEXT
5025 COLOR=0:HLIN 37,38 AT 1:HLIN 36,39 AT 2:
     HLIN 37,38 AT 3:COLOR=DC%
5030 SX=0: SY=0: N=0
5040 N = N + 1: IF N > 20 THEN N = 20  GOTO
     5170: REM LOOP
5050     GOSUB 1000: IF F2% = 1 THEN RETURN
5060     GOSUB 1200: IF F1% = 1 THEN 5050
5070     P%(N) = A: Q%(N) = B
5080     IF PC% = 1 AND D%(P%(N), Q%(N))<>0
         THEN 5100
5090     PLOT P%(N), Q%(N)
5100     IF P%(N) = P%(1) AND Q%(N) = Q%(1)
         AND N<>1 THEN 5120
5110 GOTO 5040
5120 N = N - 1
5130 FOR I=1 TO N
5140     SX = SX + P%(I): SY = SY +Q%(I)
5150 NEXT
5160 CX = SX / N: CY = SY / N: REM CALCULATE
     CENTER
5170 IF N=1 OR N=2 OR N=20 THEN FOR I=1 TO N:
     COLOR =  D%(P%(I),Q%(I)):   PLOT
     P%(I),Q%(I): NEXT: GOTO 5370
5180 IF PC%=1 AND D%(CX,CY)<>0 THEN 5200
5190 PLOT CX,CY: D%(CX,CY)=DC%: REM PLOT THE
     CENTER
5200 FOR SC = R1% TO R2 STEP R3
5210     FOR I = R1% TO N
5220         XB = (P%(I) - CX) * SC + CX
5230         YB = (Q%(I) - CY) * SC + CY
5240         XE = (P%(I + 1) - CX) * SC + CX
5250         YE = (Q%(I + 1) - CY) * SC + CY
5260         A = XE - XB: B = YE - YB
5270         L = SQR(A^2 + B^2): IF L=0 THEN
             5350
5280         UX = A / L: UY = B / L
5290         FOR J=R TO L: REM PLOT THE LINE
5300             A1% = .5 + XB + J * UX
```

```
5310                B1% = .5 + YB + J * UY
5320                IF PC% = 1 AND
                    D%(A1%,B1%)<>0 THEN 5340
5330                PLOT A1%,B1%: D%(A1%,B1%)
                    = DC%
5340                    NEXT
5350            NEXT
5360 NEXT
5370 COLOR = 15:   FOR II=1 TO 3:   HLIN 36,39
     AT II: NEXT
5380 COLOR = 0:    PLOT 38,2:   COLOR=DC%
5390 RETURN
```

The loop of lines 5040–5110 senses the X,Y values for each vertex of the polygon. An exit from the loop occurs if the number of vertex points exceeds 20 or if the stylus is pressed at the starting point. The loop may also be exited by pressing the stylus at any menu box. The loop of lines 5130–5150 calculates SX,SY. Line 5160 calculates the coordinates of the center of the polygon. Line 5170 aborts the utility if an illegal data entry has been made. Lines 5180 and 5190 plot the center of the polygon, if it is not located in a protected area. The loop of lines 5200–5360 changes the scale of the polygon and uses the same principles as the line utility does to plot the data.

Typical polygons created by means of the polygon utility are shown in figure 6-7.

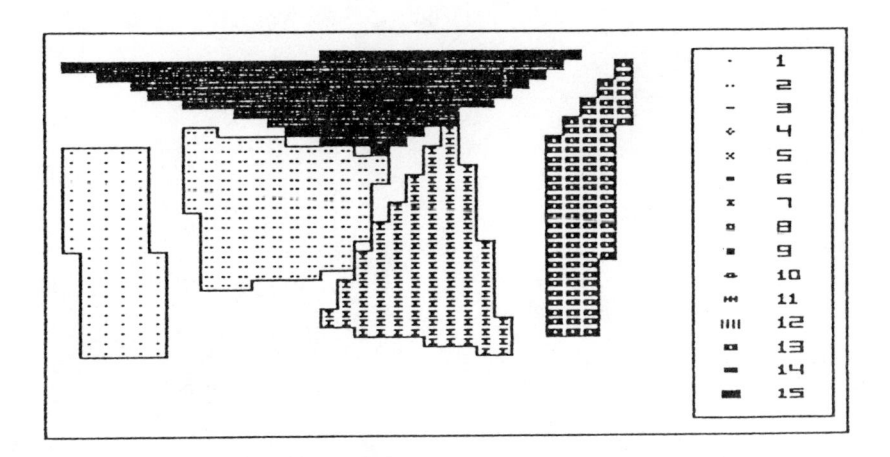

Figure 6-7. Map showing output derived from the polygon utility.

CREATION OF A MAP

The four utilities described above allow the user to enter data representing spot entities, linear and nonlinear entities, and regular- or irregular-shaped polygonal entities. The procedure for creating a map on the basis of these utilities entails three steps.

First, the basic parameters of the drawing must be determined. These parameters consist of the scale of the map, and the actual land area corresponding to one cell. Determining either parameter requires visualizing the input source map schematically as a mosaic of rectangular windows. Each window must be indexed according to its relative position in the mosaic. The size of each window in the mosaic will correspond exactly to the size of the active work space of the template. The window-sized section of the input map becomes the input source map for data entry (see figure 6-8).

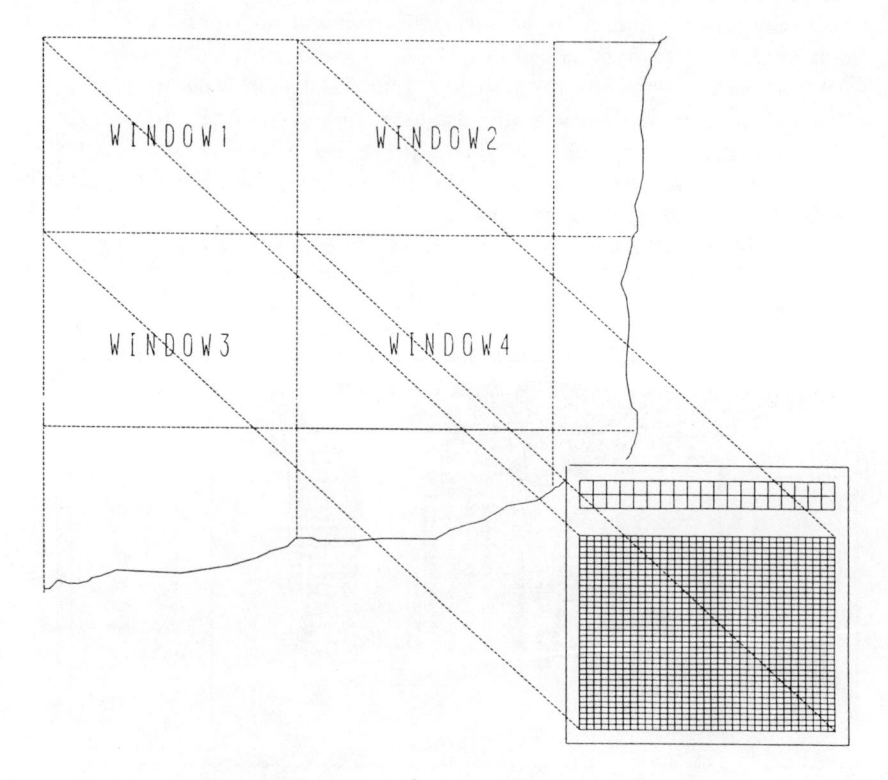

Figure 6-8. A typical mosaic showing windows.

Creating and Editing a Thematic Map

The window logically determines the linear and areal scale of the digitized map. For example, the active work space of the template may be 10 by 5¾ inches, and a 4 inches = 1 mile map may serve as the primary source map. At this point, regardless of the overall size of the primary source map, a 10- by 5¾-inch segment of the primary source map becomes the input source map for purposes of data entry. This represents a physical area of 2.5 miles by 1.4375 miles, corresponding to an area of 2,300 acres or 3.6 square miles.

Since the input source map is divided into 1,296 rectangles in a 36 by 36 matrix of cells, each cell is equivalent to 1.7746 acres. Each cell is rectangular, not square, and represents a physical area 367 feet long by 210 feet wide (a ratio of 7:4).

These figures seem unconventional as unit lengths and areas. The apparent awkwardness, however, does not affect the efficiency of computer operations or user convenience in any way. In fact, a user who gets accustomed to the data entry and interpretation process may grow to appreciate the flexibility afforded by the use of such unconventional scales.

The second step in map creation involves assigning color to represent the characteristic of the spatial entity to be digitized.

Some simple rules govern the selection of colors. Even though there are sixteen colors, only fifteen are actually usable because one (black) represents the background color. Hence, any color other than black may be selected to represent attributes of a spatial entity. If more than fifteen characteristics are present in an area being graphically represented, the characteristics must be reclassified so that they do not exceed fifteen.

The user should manually maintain a record of color assignments, in order to define the legend after digitizing.

The third and final step in map creation involves accessing the utility and digitizing the input source map. Some judgment must be exercised in digitizing and data entry, especially in the delineation of boundaries. A considerable likelihood of "overbounding" and "underbounding" exists, owing to the inherent imprecision of low-resolution graphics cells. Some allowances must be made in reconciling the inaccuracies in boundary delineation.

Although digitizing need not occur in any specific order, the user may find it expedient to begin by digitizing the entities that can be visualized as polygons, then proceed to the data entry of linear and nonlinear alignments, and finally end with the data entry of spot locations. In this way, the cell utility can be used to fine-tune the detail of the digitized map conveniently and quickly.

It is always wise to conduct some random code checking with the digitizing pen before terminating the digitization procedure. If the accuracy and reliability of the digitized map is confirmed through these random checks, the user may save the display onto disk (the save procedure is discussed in chapter 7). If the random checking uncovers problems, the map should be subjected to editing. Errors may be corrected very rapidly by reverting to the choice of colors and by selecting the appropriate editing utilities for making corrections.

In map creation, it is good practice to create and store a base map that shows the outline of the entire area being graphically represented, together with any landmarks that may be used as reference points. Thematic maps may then be digitized using the base map, with the defined boundary and the landmarks serving as fixed references. This will preclude the possibility of incorrect boundary adjustments appearing among a group of thematic maps that may be used for overlays, later on.

A little practice with the different utilities will help overcome initial uncertainties that may arise.

MAP EDITING

Map editing is necessary to correct any mistakes that may have occurred during the data entry process, to adjust boundary accuracy, or to do both. Later, map editing may also be used for updating the information on a map or for making intentional changes to reflect corresponding changes in the attributes of spatial entities.

Map editing is extremely simple. Errors in location of cells may be corrected by reentering the cell, with the correct color code at the correct location, from the graphics tablet. Alternatively, cells located incorrectly may be blanked out, by selecting the background color (black) and reentering the cell in that color.

Two special utilities, not described above, are particularly useful for editing: the cell protection utility, and the recode utility.

The Cell Protection Utility

The cell protection utility provides a means of protecting an area of predigitized cells from overwriting. If a spatial entity has been accurately digitized, the user may not wish to tamper with its boundaries accidentally while digitizing adjoining spatial entities. Without some form of cell protection, however, this may be difficult to achieve unless extreme caution is exercised during the process.

The cell protection utility enables all cells that have already been

Creating and Editing a Thematic Map

digitized to remain unaffected if the user inadvertently makes new data entries for those cells. The utility incorporates a toggle feature that activates or deactivates the utility as and when required.

The following subroutine enables a user to activate the cell protection utility:

```
6000 REM *** SUBROUTINE PROTECT ***
6010 REM THIS SUBROUTINE ACTIVATES AND
     DEACTIVATES CELL PROTECTION
6015 COLOR = 15:   HLIN 36,39 AT 5:  HLIN 36,39
     AT 6: COLOR = 0:
6020 IF PC% = 0 THEN PC% = 1: PLOT 36,6: PLOT
     37,5: PLOT 38,5: PLOT 39,6: COLOR = DC%:
     RETURN
6030 PC% = 0:   PLOT 36,5:   PLOT 37,6: PLOT
     38,6: PLOT 39,5: COLOR = DC%: RETURN
```

Lines 6020 and 6030 toggle the activation and deactivation of cell protection and indicate the current status on the status indicator. The ^ symbol on the status indicator means that cell protection is in effect; the ˇ symbol means that cell protection is not in effect. The program variable PC% serves as the flag for the cell protection utility. A value of 1 indicates that the cell protection utility is active, while a value of 0 means that this utility is inactive.

Figures 6-9 and 6-10 illustrate the difference between outputs of a drawing when protection has not (6-9) and has (6-10) been activated.

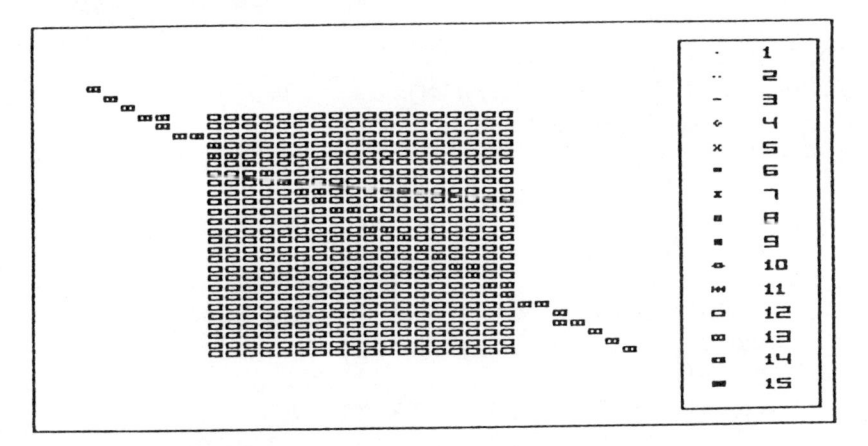

Figure 6-9. Map showing a line drawn across an unprotected polygon.

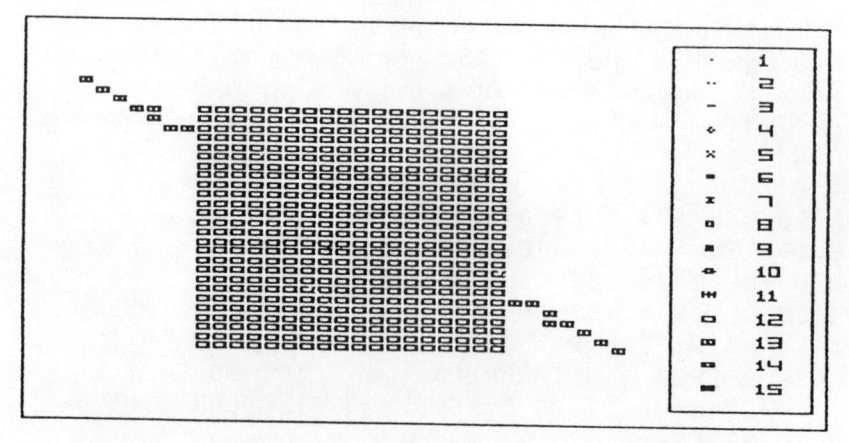

70   CARTOGRAPHY AND SITE ANALYSIS WITH MICROCOMPUTERS

Figure 6-10. Map showing a line drawn across a protected polygon.

The Recode Utility

The recode utility offers a convenient means of altering the colors (and corresponding numerical codes) of all cells displaying a particular color. If the user happens to digitize a number of spatial entities using the wrong color, it becomes very cumbersome to redigitize each of the corresponding cells individually. The recoding procedure provides a quick method for consistently altering colors.

The following subroutine enables a user to activate the recode utility:

```
7000 REM *** SUBROUTINE RECODE ***
7010 REM   THIS SUBROUTINE ENABLES COLOR
           CHANGES
7020 N=0
7030 PRINT D$;"PR#0":PRINT D$;"IN#0":   REM
     DISABLE TABLET
7040 HTAB 7:   VTAB 22:   PRINT "PRESS SPACE BAR
     TO START"
7050 HTAB 4:   VTAB 23:   PRINT "ENTER OLD CODE,
     NEW CODE ..."
7060 N = N + 1
7070 INPUT A$,B$: P%(N)  =   ABS(VAL(A$)):
     Q%(N) = ABS(VAL(B$)): IF   A$ ="" THEN
     PRINT
     PRINT D$;"PR#";SL:   PRINT "N":
     PRINT D$;"IN#";SL:
     HOME:RETURN
```

Creating and Editing a Thematic Map

```
7080 IF P%(N) <0 OR Q%(N) <0 OR P%(N) > 15 OR
     Q%(N)> 15 THEN HOME: N = N-1:
     GOTO 7040: REM DO NOT ACCEPT INPUT
7090 POKE -16368,0:  GET A$:  IF ASC(A$) <>32
     THEN HOME: GOTO 7040
7100 PRINT:   PRINT D$;"PR#";SL:   PRINT   "N":
     PRINT D$;"IN#";SL:HOME
7110 FOR J = R TO R4%
7120     FOR I = R TO R5%
7130         FOR K = R1% TO N
7140                 IF D%(I,J) = P%(K)
                     THEN
                         D%(I,J) = Q%(K):
                         COLOR = Q%(K): PLOT
                         I,J
7150                 IF D%(R4%-I,J) = P%(K)
                     THEN
                         D%(R4%-I,J) =
                         Q%(K):
                         COLOR= Q%(K): PLOT
                         R4-I,J
7160             NEXT
7170         NEXT
7180 NEXT
7190 RETURN
```

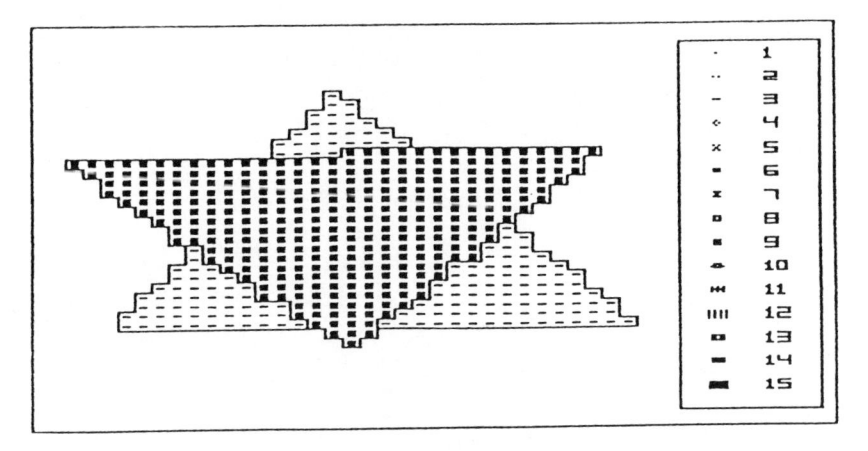

Figure 6-11. Map showing cells to be recoded.

The loop of lines 7040–7090 will be repeated until the space bar is pressed. Line 7070 seeks user-supplied data about corresponding old and new color codes. The loop of lines 7110–7180 changes the old color to the new color and updates the data matrix.

Figures 6-11 and 6-12 illustrate the results of a typical application of the recoding utility.

Figure C-2 of the color insert illustrates a typical map generated by means of the map creation and editing utilities described in this chapter.

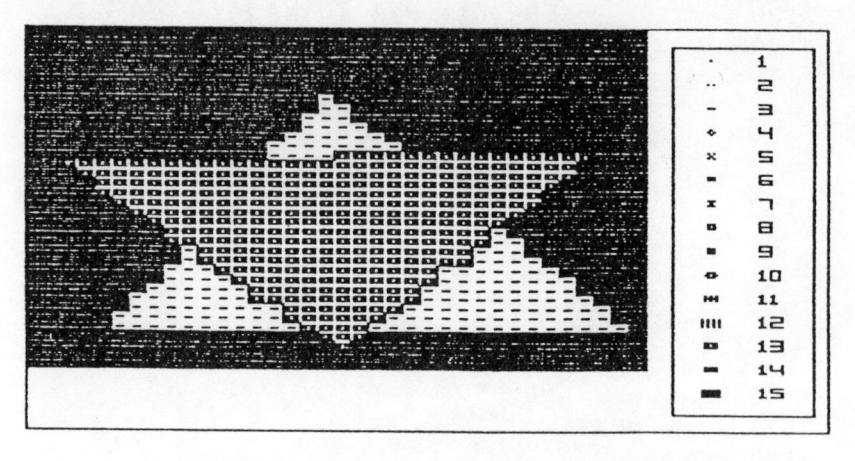

Figure 6-12. Map showing recoded cells.

SEVEN

Drawing Management Utilities

Several important drawing management utilities are required to facilitate the creation and management of a data base of graphics images. These utilities should provide the following capabilities:

- To store graphics images
- To recall and display stored graphics images
- To convert graphics images displayed on the screen into hard copy outputs
- To interpret the attributes of the graphics image's representative spatial entities (cells)
- To secure an inventory (catalog or directory) of graphics images placed in storage
- To discard unwanted graphics images from storage
- To clear the primary memory in order to start a new drawing

The utilities that perform these tasks are:

- the save utility
- the load utility
- the print utility
- the legend utility

- the catalog utility
- the delete utility
- the clear screen utility

The key to designing the save and load utilities is a sound, efficient data structure. The data structure to be used in thematic mapping is a 36 × 36 matrix of integer values. Two files are generated using this structure: a binary file for graphics images; and a text file for legends.

The save utility converts a graphics image into a matrix of numerical values and saves it as a binary file. Conversely, the load utility retrieves the binary file and reconverts its numerical values into graphics symbols and displays the image on the screen.

THE SAVE UTILITY

The save utility enables the user to save a map in a file on disk. A legend and title for the map may be created (or altered) and saved on mass storage as an additional option. The map being saved must be given a file name, by which it may be recalled for future display. The utility appends the prefix "MAP." to the file name that the user creates when the drawing is saved.

The purpose of the legend option is to enable the user to interpret the colors on the screen (graphics symbols on the hard copies) according to the attributes of the spatial entities that they represent. The legend retains the same file name as the map; however, to differentiate between the two files, the utility appends the prefix "KEY." to the legend file's file name.

The following subroutine enables a user to activate the save utility:

```
23000 REM **** SUBROUTINE SAVE ***
23010 REM THIS SUBROUTINE SAVES MAPS AND
      LEGENDS
23020 REM MAPS ARE SAVED AS BINARY FILES
23030 REM LEGENDS ARE SAVED AS TEXT FILES
23040 PRINT D$;"PR#0": PRINT D$;"IN#0"
23050 INPUT "PLEASE ENTER A NAME FOR MAP";NN$
23060 IF NN$ = "" THEN 23270
23070 GOSUB 1300
23090 N$ = "KEY." + NN$:   NN$ = "MAP."+NN$:
      REM USE PREFIX "KEY." FOR LEGEND AND
      "MAP." FOR MAP.
23100 R6% = 0 : PRINT
```

Drawing Management Utilities

```
23110 PRINT D$;"BSAVE ";NN$;
       ",A1024,L1000,D";DR$:
       REM SAVE MAP AS BINARY FILE
23120 HOME:VTAB 22: PRINT " DO YOU WISH TO
       CREATE / CHANGE LEGEND": PRINT"(Y/N) ?";
       GOSUB1400
23130 IF A$ = "N" THEN 23200
23140 HOME: INPUT "ENTER NUMBER OF ATTRIBUTES
       ?";A$
23145 R6% = VAL (A$): IF R6% >16 OR R6% < 0
       THEN 23140
23150 HOME:INPUT "ENTER TITLE FOR THE MAP
       ?";C$(17)
23160 FOR I = 1 TO R6%
23165     PRINT" ENTER COLOR CODE, ATTRIBUTE";
23170     INPUT A$,A$(I): L%(I) = VAL(A$)
23180     IF L%(I) < 0 OR L% (I) > 15 THEN
          PRINT"RANGE ERROR": GOTO 23165
23190 NEXT
23195 PRINT
23200 PRINT D$;"OPEN";N$
23210 PRINT D$;"WRITE";N$
23220 PRINT R6%: PRINT C$(17)
23230 FOR I = 1 TO R6%
23240     PRINT L%(I):PRINT A$(I)
23250 NEXT
23260 PRINT D$;"CLOSE";N$
23270 PRINT: PRINT D$;"PR#";SL: PRINT"N"
       PRINT D$;"IN#";SL: HOME
23280 RETURN
```

Line 23090 adds the "MAP." prefix to the map file name and the "KEY." prefix to the legend file name. Line 23110 saves the map image as a binary file. Line 23150 and the loop of lines 23160–23190 make possible the creation of the legend for the map. The loop of lines 23230–23250 saves the legend of the map as a text file.

THE LOAD UTILITY

The load utility allows retrieval and display of a map and its legend from mass storage. The load utility erases the current display on the screen before loading the new image.

The following subroutine enables a user to activate the load utility:

```
22000 REM *** SUBROUTINE LOAD ***
22010 REM THIS SUBROUTINE LOADS MAPS
22020 PRINT D$;"PR#0": PRINT D$;"IN#0"
22030 INPUT"PLEASE ENTER THE NAME OF THE MAP
      ?";NN$
22040 IF NN$ = "" THEN 22220
22050 GOSUB 1300
22070 N$ = "KEY." + NN$: NN$ = "MAP." + NN$
22080 PRINT: PRINT D$;"BLOAD ";NN$;
      ",A1024,D";DR$
22090 FOR J = R TO R4%
22100     FOR I = R TO R4%
22110         D%(I,J) = SCRN (I,J): REM
              CONVERT COLOR IMAGE TO NUMERIC
              DATA
22120     NEXT
22130 NEXT
22140 FOR I = 1 TO 16: L%(I) = 0: A$(I) = "":
      NEXT: C$(17) = ""
22150 PRINT: PRINT D$;"OPEN";NN$
22160 PRINT D$;"READ";NN$
22165 INPUT R6%: IF R6% = 0 THEN 22210
22170 INPUT C$(17)
22180 FOR I = 1 TO R6%
22190     INPUT L%(I): INPUT A$(I)
22200 NEXT
22210 PRINT D$;"CLOSE";NN$
22215 XM = DC%: GOSUB 21000
22220 PRINT: HOME
22225 PRINT D$;"PR#";SL: PRINT "N":
      PRINT D$;"IN#";SL
22230 RETURN
```

The load subroutine is very similar to the save subroutine. The only new part in this subroutine is the loop of lines 22090–22130, which converts color images into numerical code. The color conversion occurs in line 22110 by means of the Applesoft command SCRN, as shown. The rest of the program is self-explanatory.

THE PRINT UTILITY

The print utility is designed to produce hard copies of images that have been generated on the screen. Producing hard copies of low-resolution graphics images is more complex than accomplishing the same task for high-resolution graphics images. In the latter instance, printing involves a straightforward "screen dump" process. In the former, the low-resolution graphics cell and its color must be translated into a corresponding graphics cell and its color must be translated into a corresponding graphics symbol in high-resolution graphics mode, and then displayed on the screen. The graphics symbols can then be "screen dumped" to produce hard copy. The hard copy matches graphics symbols with corresponding numerical codes, enabling the user to interpret the attributes represented on the printed map. Reduced or enlarged hard copies can also be obtained by means of the print utility.

The following subroutine enables a user to activate the print utility:

```
18000 REM *** SUBROUTINE PRINT ***
18010 REM THIS SUBROUTINE RUNS PRINTER DRIVER
18020 PRINT D$;"PR#0": PRINT D$;"IN#0"
18030 PRINT D$;"RUN HARD COPY,D1"
18040 STOP : REM THIS LINE IS NEVER EXECUTED!
```

The print utility accesses and executes a program called HARD COPY, which is independent of the main program. The HARD COPY program converts low-resolution color graphic cells into high-resolution symbols, and then prints the latter. The HARD COPY program is written as follows:

```
1 REM *** HARD COPY PROGRAM ***
2 REM        BY A. BAHRAMI
3 REM    THIS PROGRAM CONVERTS LOW-RESOLUTION
             COLOR
4 REM    GRAPHICS TO HIGH-RESOLUTION SYMBOLS
6 REM    AND PRINTS SYMBOLS AND LEGENDS.
7 REM
8 REM
10 GOSUB 900: REM INITIALIZE VARIABLE
20 GOSUB 2000: REM GET PRINTER VARIABLE
30 REM LOOP FOR EVER
40    TEXT:GR: REM SET LOW-RESOLUTION
```

```
50      GOSUB 1000: REM LOAD FILE
60      GOSUB 3000: REM DISPLAY MENU ON SCREEN
70      HGR2: HCOLOR = 7:   REM SET SECOND PAGE OF
        HIGH-RESOLUTION GRAPHICS
80      GOSUB 4000: REM DRAW BORDER AND LEGEND
90          FOR J = R + 1 TO R4% + 1
100             Y = J * 4
110             FOR I = R TO R4%
120                 X = I * 6
130                 ON DC%(I,J) + 1 GOSUB
                    190,200,210,220,230,240,
                    250,260,270,280,290,300,
                    310,320,330,340
135                 REM DEPENDING ON COLOR OF
                    EACH CELL CALL APPROPRIATE
                    SUBROUTINE
140             NEXT
150         NEXT
160     IF MT = 1 THEN GOSUB 5000:   REM DRAW
        BORDER LINE
170     GOSUB 6000: REM PRINT
180 GOTO 30: REM END LOOP
185 END

187 REM
188 REM *** SUBROUTINES FOR COLOR CONVERSION
    ***
189 REM
190 REM *** PLOT SYMBOL FOR BLACK COLOR ***
195 RETURN
200 REM *** PLOT SYMBOL FOR MAGENTA COLOR ***
205 HPLOT X + 3,Y - 2: RETURN
210 REM *** PLOT SYMBOL FOR DARK BLUE COLOR
    ***
215 HPLOT X + 2,Y - 2: HPLOT X + 4, Y - 2:
    RETURN
220 REM *** PLOT SYMBOL FOR PURPLE COLOR ***
225 HPLOT X + 2, Y - 2 TO X + 4, Y - 2: RETURN
230 REM *** PLOT SYMBOL FOR DARK GREEN COLOR
    ***
235 HPLOT X + 3, Y - 1: HPLOT X + 2, Y - 2:
```

```
        HPLOT X + 3, Y - 3: HPLOT X + 4, Y - 2:
        RETURN
240 REM *** PLOT SYMBOL FOR GRAY 1 COLOR ***
242 HPLOT X + 2,Y - 1: HPLOT X + 2,Y - 3
244 HPLOT X + 3, Y - 2:   HPLOT X + 4,Y - 1
246 HPLOT X + 4, Y - 3
248 RETURN
250 REM *** PLOT SYMBOL FOR MEDIUM BLUE COLOR
    ***
252 HPLOT X + 2,Y - 2 TO X + 4, Y - 2
254 HPLOT X + 2,Y - 3 TO X + 4, Y - 3
256 RETURN
260 REM *** PLOT SYMBOL FOR LIGHT BLUE COLOR
    ***
262 HPLOT X + 2,Y - 1 TO X + 4,Y - 1
264 HPLOT X + 3,Y - 2
266 HPLOT X + 2,Y - 3 TO X + 4, Y - 3
268 RETURN
270 REM *** PLOT SYMBOL FOR BROWN COLOR ***
272 HPLOT X + 2,Y - 1 TO X + 4,Y - 1
274 HPLOT X + 2,Y - 2: HPLOT X + 4,Y - 2
276 HPLOT X + 2,Y - 3 TO X + 4,Y - 3
278 RETURN
280 REM *** PLOT SYMBOL FOR ORANGE COLOR ***
282 HPLOT X + 2,Y - 1 TO X + 4,Y - 1
284 HPLOT X + 2,Y - 2: HPLOT X + 4,Y - 2
286 HPLOT X + 2,Y - 3 TO X + 4,Y - 3
288 HPLOT X + 3,Y - 2: RETURN
290 REM *** PLOT SYMBOL FOR GRAY 2 COLOR ***
292 HPLOT X + 2,Y - 1 TO X + 4,Y - 1
294 HPLOT X + 2,Y - 2: HPLOT X + 4,Y - 2
296 HPLOT X + 2,Y - 3 TO X + 4,Y - 3
298 HPLOT X + 1,Y - 2: HPLOT X + 5,Y - 2:
    RETURN
300 REM *** PLOT SYMBOL FOR PINK COLOR ***
302 HPLOT X + 1,Y - 2 TO X + 5,Y - 2
304 HPLOT X + 1,Y - 1: HPLOT X + 3,Y - 1
306 HPLOT X + 5,Y - 1: HPLOT X + 1,Y - 3
308 HPLOT X + 3,Y - 3: HPLOT X + 5,Y - 3:
    RETURN
310 REM *** PLOT SYMBOL FOR GREEN COLOR ***
```

```
311 REM TWO GRAY LEVELS ARE AVAILABLE
312 IF MT = 1 THEN 316
314 HPLOT X + 1,Y - 1 TO X + 5,Y - 1:
    HPLOT X + 1,Y - 3 TO X + 5,Y - 3:
    HPLOT X + 1,Y - 2: HPLOT X + 5,Y - 2
315 RETURN
316 HCOLOR = 5: HPLOT X,Y - 1 TO X + 6,Y - 1:
    HPLOT X,Y - 3 TO X + 6,Y - 3
318 HPLOT X,Y - 2 TO X + 6,Y - 2:
    HPLOT X,Y TO X + 6, Y: HCOLOR = 7: RETURN
320 REM *** PLOT SYMBOL FOR YELLOW COLOR ***
321 REM TWO GRAY LEVELS ARE AVAILABLE
322 IF MT = 1 THEN 326
324 HPLOT X + 1,Y - 1 TO X + 5,Y - 1:
    HPLOT X + 1,Y - 3 TO X + 5,Y - 3:
    HPLOT X + 1,Y - 2: HPLOT X + 5, Y - 2
325 HPLOT X + 3,Y - 2: RETURN
326 HPLOT X + 1,Y - 1 TO X + 5,Y - 1:
    HPLOT X + 1,Y - 3 TO X + 5,Y - 3
328 HPLOT X + 1,Y - 2: HPLOT X + 5,Y - 2
    HPLOT X + 2,Y - 2: HPLOT X + 4,Y - 2:
    RETURN
330 REM *** PLOT SYMBOL FOR AQUA COLOR ***
331 REM TWO GRAY LEVELS ARE AVAILABLE
332 IF MT = 1 THEN 336
334 HPLOT X + 1,Y - 1 TO X + 5,Y - 1:
    HPLOT X + 1,Y - 3 TO X + 5,Y - 3:
    HPLOT X + 1,Y - 2: HPLOT X + 5,Y - 2
335 HPLOT X + 2,Y - 2: HPLOT X + 4,Y - 2:
    RETURN
336 HPLOT X + 1,Y - 1 TO X + 5,Y - 1:
    HPLOT X + 1,Y - 3 TO X + 5,Y - 3:
    HPLOT X + 1,Y - 2 TO X + 5,Y - 2:
338 RETURN
340 REM *** PLOT SYMBOL FOR WHITE COLOR ***
341 REM TWO GRAY LEVELS ARE AVAILABLE
342 IF MT = 1 THEN 346
344 HPLOT X + 1,Y - 1 TO X + 5,Y - 1:
    HPLOT X + 1,Y - 3 TO X + 5,Y - 3
345 HPLOT X + 1,Y - 2 TO X + 5,Y - 2: RETURN
346 HPLOT X,Y - 1 TO X + 6,Y - 1:
```

Drawing Management Utilities

```
    HPLOT X,Y - 3 TO X + 6,Y - 3:
    HPLOT X,Y - 2 TO X + 6,Y - 2
348 HPLOT X,Y TO X + 6,Y: RETURN
349 REM
350 REM *** SUBROUTINES TO PLOT NUMBER 1 - 15
    ***
351 REM
390 REM REM *** PLOT 0 ***
392 HPLOT X,Y TO X + 3,Y TO X + 3,Y - 4 TO X,Y
    - 4 TO X,Y
398 RETURN
400 REM *** PLOT 1 ***
402 HPLOT X + 1,Y TO X + 1,Y - 4
404 HPLOT X,Y - 3: HPLOT X,Y TO X + 2,Y
408 RETURN
410 REM *** PLOT 2 ***
412 HPLOT X + 4,Y TO X,Y TO X,Y - 2 TO
    X + 4,Y - 2 TO X + 4,Y - 4 TO X,Y - 4
418 RETURN
420 REM *** PLOT 3 ***
422 HPLOT X,Y TO X + 4,Y TO X + 4,Y - 2 TO X,Y
    - 2
424 HPLOT X,Y - 4 TO X + 4,Y - 4 TO X + 4,Y -
    2
428 RETURN
430 REM *** PLOT 4 ***
432 HPLOT X,Y - 4 TO X,Y - 2 TO X + 4,Y - 2:
    HPLOT X + 4,Y TO X + 4,Y - 4
438 RETURN
440 REM *** PLOT 5 ***
442 HPLOT X,Y TO X + 4,Y TO X + 4,Y - 2 TO X,Y
    - 2 TO X,Y - 4 TO X + 4,Y - 4
448 RETURN
450 REM *** PLOT 6 ***
452 HPLOT X + 4,Y - 4 TO X,Y - 4 TO X,Y TO X +
    4,Y TO X + 4,Y - 2 TO X,Y - 2
458 RETURN
460 REM *** PLOT 7 ***
462 HPLOT X + 4,Y TO X + 4,Y - 4 TO X,Y - 4
468 RETURN
470 REM *** PLOT 8 ***
```

```
472 HPLOT X,Y TO X + 4,Y TO X + 4,Y - 2 TO X,Y
    - 2 TO X,Y - 4 TO X + 4,Y - 4 TO X + 4,Y -
    2
476 HPLOT X,Y - 4 TO X,Y - 2:
    HPLOT X,Y TO X,Y - 2
478 RETURN
480 REM *** PLOT 9 ***
482 HPLOT X,Y TO X + 4,Y TO X + 4,Y - 2 TO X,Y
    - 2 TO X,Y - 4 TO X + 4,Y - 4 TO X + 4,Y -
    2
484 HPLOT X,Y - 4 TO X,Y - 2
486 RETURN
490 REM *** PLOT 10 ***
492 GOSUB 400: REM PLOT 1
494 X = X + 4
496 GOSUB 390: REM PLOT 0
498 RETURN
500 REM *** PLOT 11 ***
502 GOSUB 400: REM PLOT 1
504 X = X + 4
506 GOSUB 400: REM PLOT 1
508 RETURN
510 REM *** PLOT 12 ***
512 GOSUB 400: REM PLOT 1
514 X = X + 4
516 GOSUB 410: REM PLOT 2
518 RETURN
520 REM *** PLOT 13 ***
522 GOSUB 400: REM PLOT 1
524 X = X + 4
526 GOSUB 420: REM PLOT 3
528 RETURN
530 REM *** PLOT 14 ***
532 GOSUB 400: REM PLOT 1
534 X = X + 4
536 GOSUB 430: REM PLOT 4
538 RETURN
540 REM *** PLOT 15 ***
542 GOSUB 400: REM PLOT 1
544 X = X + 4
546 GOSUB 440: REM PLOT 5
548 RETURN
```

```
900 REM *** INITIALIZE VARIABLE ***
910 R4% = 35: R = 0: D$ = CHR$(4)
920 DIM D%(R4%+2,R4%+2),C$(17),A$(30),
    L%(16),Q%(15)
960 RETURN
1000 REM *** SUBROUTINE 1000 ***
1002 REM THIS SUBROUTINE LOADS MAP INTO MEMORY
1004  REM IN CASE OF NULL ENTRY, THE MAIN
     PROGRAM RUNS
1010 INPUT "PLEASE ENTER THE NAME OF THE MAP
     ?";NN$
1020 IF NN$ = "" THEN PRINT :
     PRINT D$;"RUN MAIN,D1"
1030 PRINT "DISK DRIVE # 1 OR 2 ?";:GET DR$:
     PRINT
1040 N$ = "KEY." + NN$: NN$ = "MAP." + NN$
1050 PRINT: PRINT D$;"BLOAD";NN$;
     ",A1024,D",DR$
1060 FOR J = R TO R4%
1070      FOR I = R TO R4%
1080           D%(I,J) = SCRN (I,J)
1090      NEXT
1100 NEXT
1110 PRINT: HOME
1120 PRINT D$;"OPEN";N$: REM READ LEGEND FILE
1130 PRINT D$;"READ";N$
1140 INPUT R6%
1150 IF R6% = 1 THEN 1200
1160 INPUT C$(17)
1170 FOR I = 1 TO R6%
1180      INPUT L%(I): INPUT A$(I)
1190 NEXT
1200 PRINT D$;"CLOSE";N$
1210 HOME: RETURN
2000 REM *** SUBROUTINE 2000 ***
2010 REM THIS SUBROUTINE GETS PRINTER VARIABLE
2020 TEXT: HOME
2030 INPUT "PRINTER SLOT NUMBER (DEFAULT=1) ";
     PR$
2040 IF PR$ = "" THEN PR$ = "1"
2050 INPUT "PRINTER COMMAND ";CO$
2060 INPUT " ANY CHANGES ?"; GET A$
```

```
2070 IF A$ = "Y" THEN 2020
2080 HOME:RETURN
3000 REM *** SUBROUTINE 3000 ***
3010 REM THIS SUBROUTINE DISPLAYS MENU
3020 TEXT: HOME
3025 HTAB 15: VTAB 1: PRINT "PRINT MENU"
3030 VTAB 5: HTAB 5: PRINT "(1) MAP WITH
     BORDER GRAY LEVEL-1"
3040 VTAB 10: HTAB 5: PRINT "(2) MAP WITHOUT
     BORDER GRAY LEVEL-2"
3050 VTAB 20: HTAB 5: PRINT "PICK SELECTION
     ?";
3060 POKE -16368,0: GET A$
3070 IF A$<> "1" AND A$ <> "2" THEN 3060
3075 MT = VAL (A$): PRINT
3200 HOME: RETURN
4000 REM *** SUBROUTINE 4000 ***
4010 REM THIS SUBROUTINE DISPLAYS BORDER AND
     LEGEND ON THE SCREEN
4020 HPLOT 225,8 TO 274,8 TO 274,163 TO
     225,163 TO 225,8
4030 HPLOT 0,0 TO 279,0 TO 279,170 TO 0,170 TO
     0,0
4040 K = 0
4050 FOR Y = 15 TO 155 STEP 10
4060     K = K + 1
4070     X = 255 REM PLOT 1 - 15
4080     ON K GOSUB 400,410,420,430,440,450,
             460,470,480,490,500,510,520,530,540
4090     X = 235: REM PLOT SYMBOLS
4100     ON K GOSUB 200,210,220,230,240,250,
             260,270,280,290,300,310,320,330,340
4110 NEXT
4200 RETURN
5000 REM *** SUBROUTINE 5000 ***
5010 REM THIS SUBROUTINE DRAWS BORDER LINE
5020 FOR J = R + 1 TO R4% + 1
5030     FOR I = R TO R4%
5040         Y = J * 4: X = I * 6
5050         IF D%(I,J) <> D%(I,J + 1)
                 THEN HPLOT X,Y TO X + 6,Y
5060         IF D%(I,J) <> D%(I + 1,J)
```

Drawing Management Utilities 85

```
                       THEN HPLOT X + 6,Y - 4 TO
                       X + 6,Y
5070          NEXT
5080 NEXT
5200 RETURN
6000 REM *** SUBROUTINE 6000 ***
6010 REM THIS SUBROUTINE PRINTS THE MAP AND
     LEGEND
6020 PRINT
6030 PRINT D$;"PR#";PR$: REM PRINTER SLOT
6040 PRINT CO$: REM PRINTER COMMAND
6060 IF R6% = 0 THEN 6150
6070 PRINT "#     KEY": PRINT
6080 FOR I = 1 TO R6%
6090      PRINT L%(I);:HTAB 3: PRINT "=";
6100      PRINT A$(I)
6110 NEXT
6120 PRINT: PRINT
6130 PRINT C$(17)
6140 PRINT D$;"PR#0"
6150 HOME: RETURN
```

The HARD COPY program listed above can be viewed in two parts: the subroutines, and the main body.

Subroutines

Sixteen subroutines are contained between lines 190 and 340. Each subroutine converts a color in the low-resolution graphics mode into a unique symbol in the high-resolution graphics mode.

Sixteen text symbols, representing the numbers 0 to 15, are also designed in the high-resolution graphics mode for use in the legend. Sixteen subroutines for performing this task are embedded between lines 390 and 540.

Subroutine 1000 loads the low-resolution map and legend into memory. In the event of a null entry on line 1010, the subroutine runs the main HARD COPY program.

Subroutine 2000 seeks information relating to the printer, such as slot number and printer command control string for dumping a graphics page from the screen onto the printer. Command control strings vary from one printer to another; this information may be obtained from the support documentation supplied by the printer's manufacturer.

Subroutine 3000 displays a menu of display options on the screen.

86   CARTOGRAPHY AND SITE ANALYSIS WITH MICROCOMPUTERS

The user may opt for a map with or without a border, and with symbols printed in either of two gray levels.

Subroutine 4000 draws the border of the map. The loop of lines 4050–4110 draws the legend on the right side of the screen, by calling the appropriate subroutines discussed above.

Subroutine 5000 dumps the screen image onto the printer.

The Main Body

Line 10 calls subroutine 900, which initializes variables. Line 20 calls subroutine 2000, to obtain all the printer variables. The loop of lines 30–180 is perpetual; it can only be exited by making a null entry in subroutine 1000.

After the map has been loaded and the map customization option has been selected, line 80 calls subroutine 4000 to draw the border and legend. The loop of lines 90–150 converts low-resolution graphics cells into 6 × 4 dot matrix symbols and calls the conversion subroutines on the basis of the value (color code) of matrix D%. Line 170 calls subroutine 6000 to print the display. Line 180 repeats the process.

Figures 7-1 and 7-2 illustrate sample printouts of the color display depicted in figure C-2 of the color insert.

THE LEGEND UTILITY

The legend utility facilitates quick computations and display of statistical profiles of spatial information from a map. These statistical profiles primarily take the form of frequency distributions of different attributes that are characterized and displayed.

The statistical profiles may be generated for any map by invoking the legend utility module. The utility seeks out the source map, determines the frequency of occurrence of each attribute among the various cells in the display, and returns a report of frequency distribution.

This capability is extremely useful in the preparation and compilation of statistical profiles of land use, density, vegetation, distribution of resources, and the like. It is invaluable in preparing and compiling visual displays and statistical profiles of information for census tracts, covering such items as income, ethnicity, and employment.

The following subroutine enables a user to activate the legend utility:

```
17000 REM **** SUBROUTINE LEGEND ***
17010   REM THIS SUBROUTINE DISPLAYS LEGENDS ON
        THE SCREEN AND CONDUCTS FREQUENCY
        COUNTS
```

Drawing Management Utilities 87

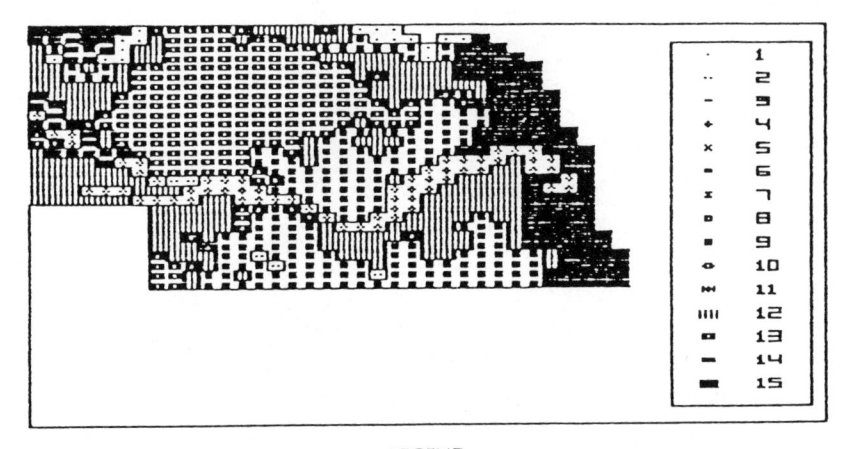

LEGEND

#	Key	Freq.
1	Bluffs & Escarpments	17
2	Large Reservoirs	5
4	Valleys	62
9	Dissected Plains	190
12	Plains	201
13	Sand Hills	211
14	Valley-side Slopes	26
15	Rolling Hills	113

Figure 7-1. Hard copy of the color display in figure C-2.

```
17020 PRINT D$;"PR#0": PRINT D$;"IN#0"
17030 IF R6% = 0 AND C$(17) = " " THEN
      PRINT"LEGEND DOES NOT EXIST":
      FOR II = 1 TO 500: NEXT: GOTO 17270
17040 HOME:  VTAB 23:  PRINT"DO YOU WANT
      FREQUENCY (Y/N) ? ";: GOSUB 1400
17060 IF A$ = "N" THEN 17160
17070 HOME: VTAB 23: HTAB 12:
      PRINT" << CALCULATING >>"
17080 FOR I = 0 TO 15: Q%(I) = 0: NEXT
17090 FOR I = R TO R4%
17100     FOR J = R TO R4%
17110         FOR K = 1 TO R6%
17120             IF L%(K) = D%(I,J) THEN
                  Q%(L%(K)) = Q%(L%(K))+1
17130         NEXT
17140     NEXT
```

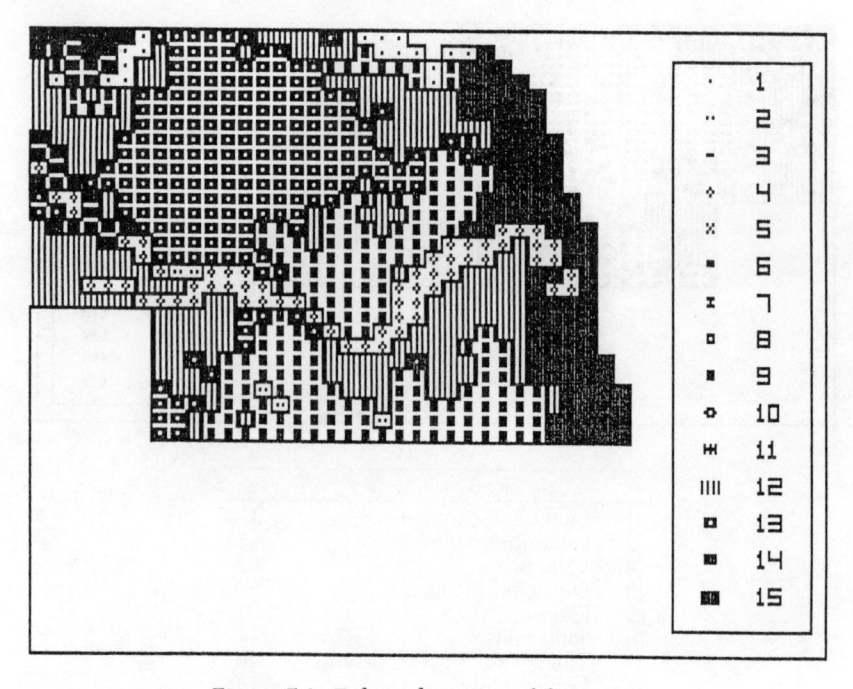

88   CARTOGRAPHY AND SITE ANALYSIS WITH MICROCOMPUTERS

Figure 7-2. *Enlarged version of figure 7-1.*

```
17150 NEXT
17160 TEXT: HOME: INVERSE: HTAB 11: VTAB 1:
      PRINT"    LEGEND    "
17170 IF A$ = "Y" THEN
      PRINT"L SYMBOLS      KEY
      FREQ": GOTO 17190
17180 PRINT"L SYMBOLS      KEY"
17190 NORMAL: PRINT
17200 FOR I = 1 TO R6%
17210     PRINT L%(I);: HTAB 3: PRINT C$(I);:
          HTAB 14: PRINT "=";:
          PRINT A$(I);:HTAB 37
17220     IF A$ = "N" THEN PRINT: GOTO 17240
17230     PRINT Q%(L%(I)): REM PRINT FREQUENCY
17240 NEXT
17250 HTAB 2: VTAB 22:PRINT C$(17)
17260 HTAB   10:   VTAB 24:
      PRINT"<< PRESS ANY KEY >>";: GET A$
```

```
17265 HOME: GR: XM = DC% : GOSUB 21000: GOSUB
      970
17270 PRINT: HOME
17280 PRINT D$;"PR#";SL: PRINT"N":
      PRINT D$;"IN#";SL
17290 RETURN
```

The legend subroutine displays on the screen a legend of attributes and their frequencies. The frequency count is set up as an option because the process is relatively time-consuming. Line 17030 verifies that the flag R6% is equal to zero. If this is the case, the legend does not exist and the utility will be terminated. The loop of lines 17090–17150 is executed only if the user answers "Y" to the query on frequency count. This loop counts the number of cells that share various attributes. Line 17160 switches from the low-resolution graphics screen to the text screen. The loop of lines 17200–17240 displays the legend along with other map features. Line 17265 switches from the text screen back to the low-resolution graphics screen and calls subroutine 21000 to redraw the status indicator on the screen.

THE CATALOG UTILITY

The catalog utility enables the user to obtain a display of the contents of the entire library of data files on the disk. (This utility can be replaced by a corresponding DOS utility.)

All files marked with the "MAP." prefix represent maps, whereas all files marked with the "KEY." prefix represent legend and title files.

The following subroutine enables a user to activate the catalog utility:

```
24000 REM *** SUBROUTINE CATALOG ***
24010 REM THIS SUBROUTINE DISPLAYS CATALOG OF
      FILES
24020 PRINT D$;"PR#0": PRINT D$;"IN#0"
24030 GOSUB 1300: REM GET DISK DRIVE #
24040 TEXT: HOME
24050 PRINT:PRINT D$;"CATALOG,D";DR$
24060 VTAB 24:HTAB 12:  PRINT "<<PRESS ANY
      KEY >>";: GET A$
24070 GR: XM=DC%: GOSUB 21000
24080 GOSUB 970: REM REDRAW MAP
```

```
24090 PRINT: PRINT D$;"PR#";SL:PRINT "N":
      PRINT D$; "IN#";SL: HOME
24100 RETURN
```

The subroutine is self-explanatory.

THE DELETE UTILITY

The delete utility enables the user to delete map files and legend files from mass storage, in the event that they become obsolete. It is also customarily used to open up disk space. (This utility can be replaced by a corresponding DOS utility.)

The following subroutine enables a user to activate the delete utility:

```
20000 REM *** SUBROUTINE DELETE ***
20010 REM THIS SUBROUTINE DELETES FILES FROM
      DISK
20020 PRINT D$;"PR#0": PRINT D$;"IN#0"
20030 INPUT "PLEASE ENTER THE NAME OF THE MAP
      TO BE DELETED ?";NN$
20040 IF NN$="" THEN 20100
20050 GOSUB 1300
20060 N$ = "KEY." + NN$: NN$ = "MAP." + NN$
20070 PRINT: PRINT D$;"DELETE ";NN$;",D";DR$
20080 PRINT D$;"DELETE ";N$;",D";DR$
20100 PRINT: HOME
20110 PRINT D$;"PR#";SL:PRINT D$;"IN#";SL
20120 HOME: RETURN
```

Line 20030 seeks the name of the file to be deleted; a null entry to this query terminates the delete utility. Lines 20070 and 20080 use the DOS command to delete map and legend files, respectively.

THE CLEAR SCREEN UTILITY

The clear screen utility enables the user to clear the memory of the computer and to clear the screen in order to commence creating and editing a new map. This utility is essential at the start of a new digitizing session.

The following subroutine enables a user to activate the clear screen utility:

```
19000 REM *** SUBROUTINE CLEAR ***
19010 REM THIS SUBROUTINE CLEARS SCREEN AND
      MEMORY
19020 PRINT D$;"PR#0": PRINT D$;"IN#0"
19030 VTAB 23: HTAB 10: PRINT " CONFIRM
      (Y/N)";
19040 GOSUB 1400
19050 IF A$ < > "Y" THEN 19200
19060 GR: XM = DC%: GOSUB 21000: REM CLEAR
      SCREEN
19065 FOR I = 1 TO R6%
19068       L%(I) = 0: A$(I) = " "
19069 NEXT: R6% = 0: C$(17) = " "
19070 FOR I = R TO R4%: REM CLEAR MATRIX
19080      FOR J = R TO R4%
19090           D%(I,J) = 0:E%(I,J) = 0
19100      NEXT
19110 NEXT
19200 PRINT: PRINT D$;"PR#";SL: PRINT "N":
      PRINT D$;"IN#";SL: HOME
19210 RETURN
```

This subroutine is self-explanatory.

APPLICATIONS OF DRAWING MANAGEMENT UTILITIES

Two of the most common applications of the drawing management utilities are in managing graphics data bases and in map updating.

A typical planning or design office is likely to create and maintain a data base consisting of a large assortment of thematic maps and legend files covering areas that have been studied in the past. These files may be stored on a hard disk or in a diskette library. On numerous occasions, one or more of the map or legend files will have to be retrieved and accessed for some purposes. The catalog, load, and save utilities will repeatedly be used to review files, to access and retrieve data for decision-making, and to store data for future use. The print utility may frequently be activated to retrieve hard copies of maps to accompany reports. The clear screen utility will be used constantly to clear a display whose purpose has been served or to prepare for a new display that has to be created. In all these instances, the drawing management utilities are invaluable.

Another typical need encountered in planning and design offices is that of updating data on file to reflect changes that have occurred over time. For instance, changes regularly take place in land use, topography, population density, and land value. The corresponding thematic maps must therefore be updated from time to time. If historical data are to be preserved, the updating process may have to retain both original and updated versions of data.

Several utilities are used in tandem to update maps and legends. The disk on which the map is located must first be identified by means of the catalog utility. Once it is located, the map may be retrieved and displayed by means of the load utility.

The spatial entities for which alterations or updates in attribute information are to be made must be identified and located. The color code corresponding to the altered/updated status of the entities must be selected. The editing utilities will facilitate the process of updating. The updates will override the existing data entry in each affected cell.

If one or more attributes need to be altered consistently in a map, the most convenient updating method is to use the recode utility. The selected attribute(s) can be recoded in a single step. (The recode utility must not be used if an attribute occurs at several locations on a map display but requires alteration at only selected locations among them.)

Updated files should be assigned new file names when saved onto disk. This is because, if an updated file is saved using the original file name, it will destroy and take the place of the original file stored on disk—thereby eliminating any opportunity to retrieve the original map in its unaltered state at a later date. A good practice is to maintain a series of updated files, appropriately renamed or numbered to reflect the updates that have been performed; this guarantees the availability of a historical record of all changes that have occurred in a particular series of thematic maps.

The catalog and the delete tasks may be performed directly from the disk operating system, instead of invoking the corresponding utilities mentioned above. The choice between selecting the utility or the DOS command to perform the desired function is entirely up to the user. Either option delivers identical end results.

EIGHT

Simple Scanning and Opaque Overlay Utilities

A number of simple tasks may be performed using the thematic map displays described in previous chapters. Scanning, retrieval, and display of selected attributes on a thematic map are routine applications performed by planners, designers, and landscape architects in locating a particular type of site.

THE SCAN UTILITY

There are two levels of scanning: the simple scanning process, where finding the location of desired attribute(s) requires investigation of only one thematic map; and the complex scanning process, where finding the desired attribute(s) requires both sorting through a library of thematic maps and investigation of each map.

The scanning utility offers a very convenient means of performing the first type of activity. It searches for, finds, and displays the locations of all cells of a particular color; cells of all other colors are not displayed. Identifying the color to be sought is accomplished by touching any cell displaying the desired color with the stylus.

In order to conduct a scan, the user first loads the map containing the desired attribute(s). The user may then invoke the scan utility to search for and display all cells possessing the desired attributes. All other attributes will be suppressed.

The following subroutine enables a user to activate the scan utility:

```
8000 REM *** SUBROUTINE SCAN ***
8010 REM THIS SUBROUTINE SCANS FOR SELECTED
     ATTRIBUTE(S)
8020 PRINT D$;"PR#0": PRINT D$;"IN#0"
8030 INPUT "ENTER NUMBER OF DESIRED ATTRIBUTES
     ";A$
8035 HOME
8040 N = VAL(A$): I = 1
8050 IF N = 0 THEN 8230
8060 REM LOOP
8070       PRINT D$;"PR#0":HTAB 1: VTAB 22:
           PRINT "<<PRESS STYLUS AT DESIRED
           ATTRIBUTE>>"
8090       PRINT: PRINT D$;"PR#";SL: PRINT "N"
8100       GOSUB 1000: IF F2% = 1 THEN 8230
8110       GOSUB 1200: IF F1% = 1 THEN 8100
8120       PRINT D$;"PR#0": VTAB 22: HTAB
           36: PRINT "<"I">"
8130       P%(I) = D% (A,B)
8140 I = I + 1: IF I <=N THEN 8060
8150 FOR I = R TO R4%
8160     FOR J = R TO R4%
8170         COLOR = 0: PLOT I,J: REM
             CLEAR SCREEN
8180         FOR K = 1 TO N
8185             IF D%(I,J) = P%(K) THEN
                 COLOR = D%(I,J): PLOT I,J:
                 REM IF THE COLOR EQUALS
                 THE SELECTED COLOR THEN
                 PLOT THE CELL
8190             NEXT
8200     D%(I,J) = SCRN(I,J): REM UPDATE
         MATRIX
8210     NEXT
8220 NEXT
8230 PRINT: HOME
8240 PRINT D$;"PR#";SL: PRINT "N": PRINT
     D$;"IN#";SL
8250 RETURN
```

Simple Scanning and Opaque Overlay Utilities 95

Line 8030 requests the user to specify the number of attributes that are to be scanned. The loop of lines 8060–8140 saves the numerical codes of the selected colors in array P%(I). The loop of lines 8150–8220 clears the screen, checks each cell for the selected colors, and plots on the screen only cells having the selected colors. Line 8200 updates the matrix.

Figures 8-1 and 8-2 illustrate the use of the scan utility.

The second type of scanning is best performed with the multiple-map overlay utility, described in chapter 9.

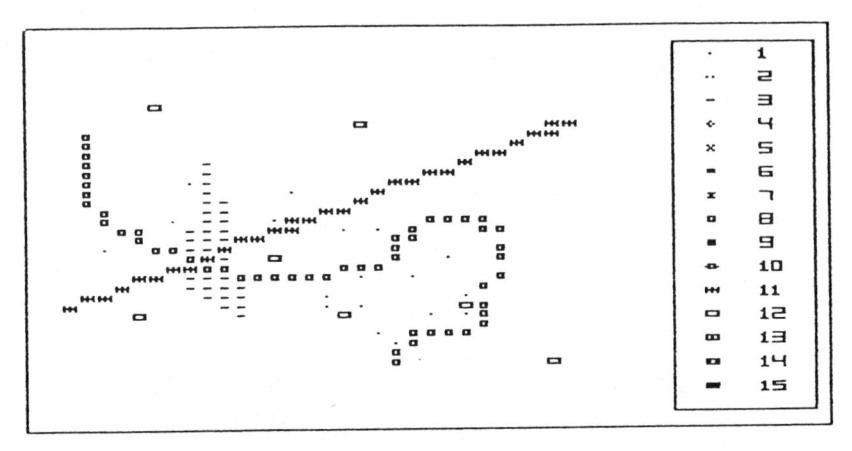

Figure 8-1. Map showing assorted attributes.

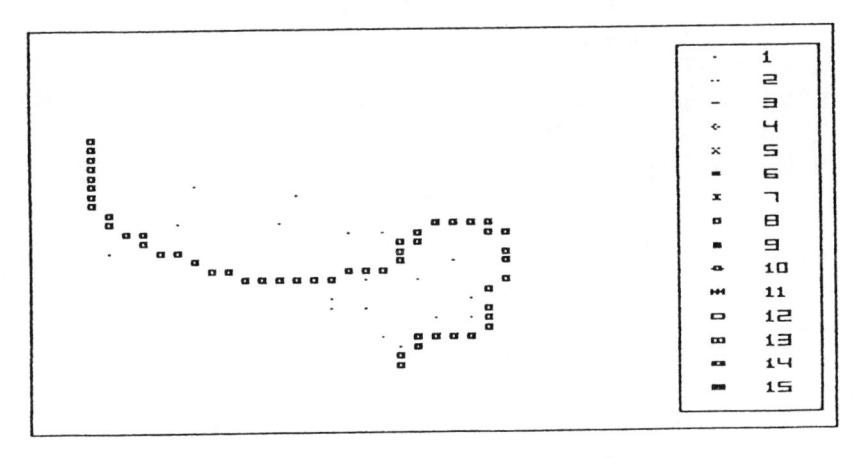

Figure 8-2. Map showing selected attributes.

THE OPAQUE OVERLAY UTILITY

The opaque overlay utility is another useful and convenient utility for planners and designers. It enables the user to superimpose the contents of one map over the contents of another.

The opaque overlay process requires two maps: The user designates one of these maps as the base map (the one upon which the opaque overlay is to be performed); the other map becomes the opaque overlay map.

The user may want to select and superimpose only certain attributes of the overlay map upon the base map. This entails selecting and designating color(s) to be rendered transparent; all other colors will be rendered opaque. The opaque overlay utility superimposes all opaque colors from the overlay map on the corresponding locations of the base map.

The following subroutine enables a user to activate the opaque overlay utility (the base map must either be created or be loaded from disk, prior to activation of the utility):

```
9000 REM *** SUBROUTINE OPAQUE OVERLAY ***
9010 REM THIS SUBROUTINE PERFORMS OPAQUE
     OVERLAYS
9020 PRINT D$;"PR#0": PRINT D$;"IN#0"
9030 INPUT "ENTER THE NAME OF OVERLAY MAP ?";
     NN$
9040 IF NN = "" THEN 9220
9050 GOSUB 1300
9060 HOME: INPUT "ENTER NUMBER OF
     TRANSPARENT COLOR(S) ?";A$
9065 N = VAL (A$)
9070 IF N = 0 THEN 9220
9080 FOR I = 1 TO N
9090        HOME:   INPUT "ENTER  COLOR  CODE
            OF TRANSPARENT COLOR(S) ?";A$
9100        P%(I) = VAL (A$)
9110 NEXT
9120 NN$ = "MAP." + NN$
9130 PRINT: PRINT D$;"BLOAD ";NN$;
     ",A1024,D";DR$
9135 HOME
9140 GOSUB  1500:  REM SAVE THE IMAGE INTO
     MATRIX E%(R4%,R4%)
9150 FOR I = R TO R4%
```

Simple Scanning and Opaque Overlay Utilities 97

```
9160            FOR J = R TO R4%
9165               COLOR = E%(I,J): PLOT I,J
9170               FOR K = 1 TO N
9180                  IF E%(I,J) = P%(K) THEN
                      COLOR = D%(I,J): PLOT I,J
9190               NEXT
9195               D%(I,J) = SCRN (I,J) :  REM
                   UPDATE THE MATRIX
9200            NEXT
9210 NEXT
9220 PRINT: HOME
9230 PRINT D$;"PR#";SL: PRINT"N": PRINT D$;
     "IN#";SL
9240 RETURN
```

The algorithm of the opaque overlay subroutine is similar to that of the scan subroutine. The only unique part is subroutine 1500, which needs to be inserted into the main body of the main program. The listing for this subroutine is as follows:

```
1500 REM *** SUBROUTINE CONVERT ***
1510 REM   THIS  SUBROUTINE  CONVERTS  COLORS
     TO NUMERICAL CODE
1520 FOR I = R TO R4%
1530    FOR J = R TO R4%
1540       E%(I,J) = SCRN (I,J)
1550    NEXT
1560 NEXT
1570 RETURN
```

Subroutine 1500 converts low-resolution graphics colors to corresponding numerical code and saves these values in the matrix E% instead of in the matrix D%.

Figure 8-3 displays a map upon which the map in figure 8-2 is to be overlaid. Figure 8-4 illustrates the result of the overlay procedure.

TYPICAL APPLICATIONS FOR THE SCAN AND OPAQUE OVERLAY UTILITIES

The scan utility is most useful to planners and designers as a means of locating sites that possess one or more specific attributes. For instance,

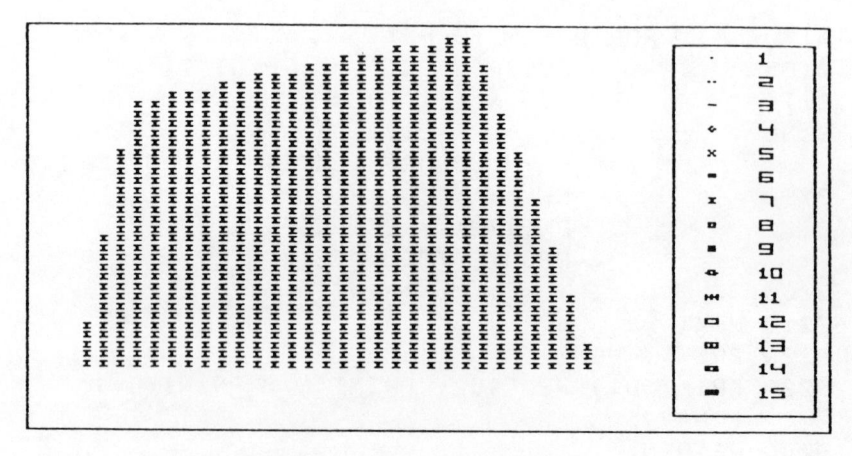

98   CARTOGRAPHY AND SITE ANALYSIS WITH MICROCOMPUTERS

Figure 8-3. Map on which figure 8-2 is to be overlaid.

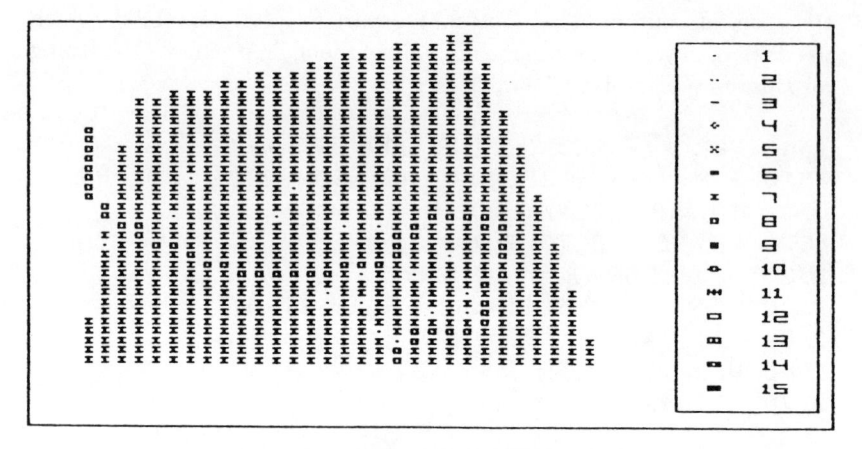

Figure 8-4. Results of the opaque overlay.

a landscape architect may use it to find a match between appropriate planting locations within a site and areas that possess specific soil type(s). The landscape architect begins by retrieving the soils map of the locality in which the site is located, by loading the appropriate drawing file. The legend of the drawing is then checked to determine the matching color code(s) for the soil type(s) being sought. The scan utility is then invoked to locate all cells painted in the color(s) being sought. The product of the scan is a map in which all cells possessing the required color(s) are displayed; display of all other cells is suppressed. The cell locations provide the landscape architect with a guideline for undertaking detailed delineation of optimum planting locations.

The scan utility is useful in a number of other applications, as well. Locations of census tracts characterized by high rates of unemployment, for example, may be identified by invoking the scan utility and applying it to maps showing the distribution of employment by census tract. In another typical application, land developers may invoke the utility to locate all sites on a floodplain map that are likely to be submerged in the event of flooding.

The primary limitation on using this utility is that all attributes being sought in the scan process must exist on one map (in combination with a number of other attributes). If the particular attributes being sought are located on different maps, the scan utility would be an ineffective way to conduct the investigation. A more appropriate way to conduct the search in that case (as noted earlier) would be to use the multiple-map overlay utility, discussed in chapter 9.

The opaque overlay utility is very useful for undertaking certain drafting tasks. In the soil types example discussed above, the product of the search process was a map showing the locations of all cells representing terrain of the desired soil type. Although the absolute locations of these cells have been identified, their locations relative to a specified landmark are as yet determinable only by conjecture.

An ideal remedy to this situation is to superimpose the map showing the desired soil type over the map showing the designated landmarks. By this means, the soil type locations and the landmarks will both be displayed in a composite map. The user should exercise discretion in interpreting the composite map, since it is neither a soils map nor a landmarks map; rather, it is a hybrid visual aid displaying the locations of a selected soil type relative to known landmarks in a given environment.

The opaque overlay utility allows construction of composite maps, but discretion is required in using this utility. It delivers effective and comprehensible results if no more than two maps are used at a time, if the same colors do not appear on both maps, and if the composite maps are used principally for visual display purposes, rather than for analysis. In all other composite mapping situations, the multiple-map overlay utility is preferable.

NINE

The Multiple-Map Overlay Utility

The general map overlay technique was developed by McHarg in his book *Design with Nature*.[1] The technique involves a mechanical process of drafting and overlaying transparencies of site maps to identify sites possessing certain desired attributes that render them suitable for a particular purpose. The transparency overlay principle is used in the multiple-map overlay utility described in this chapter.

DIFFERENCE BETWEEN MULTIPLE-MAP OVERLAYING AND OPAQUE OVERLAYING

The multiple-map overlay utility enables the user to perform logical overlays of thematic maps previously generated by the map creation and editing utilities. Whereas the opaque overlay process corresponds to a physical overlay (and may be regarded merely as a graphic display enhancement process), the multiple-map overlay is an analytical process involving the systematic weighting and rating of desired attributes and the displaying of their locations. The logical overlay process is therefore more appropriately considered to be a sophisticated enhancement of the scanning process, discussed in chapter 8.

FEATURES OF THE MULTIPLE-MAP OVERLAY UTILITY

The multiple-map overlay utility permits the logical overlay of an unlimited number of thematic maps simultaneously, with each map containing up to sixteen attributes. It allows the user to assign relative weights to each group of attributes and desirability ratings to each of the sixteen (or fewer) attributes that are displayed on each thematic map.

The multiple-map overlay is derived from the following mathematical formulas:

$$O_{ij} = (1/W) \times \sum_{k=1}^{n} A_{ijk}$$

$$W = \sum_{k=1}^{n} w_k$$

where:

O_{ij} = color code of the cell in row i, column j of the overlay product map
n = number of maps to be overlaid
w_k = weight factor for the kth map
W = summation of all weight factors
A_{ijk} = color code of the cell in row i, column j of map k

WORKING PROCEDURE OF THE MULTIPLE-MAP OVERLAY UTILITY

The multiple-map overlay utility consists of two procedures: the first involves tasks performed off-line by the user; the second involves tasks performed on-line by the computer.

Off-Line Procedure

The initial step in the multiple-map overlay procedure is to assemble all the maps that are to be used. Only maps that contain one or more attributes in need of representation should be selected. Next, the user must identify and specify the attributes in each map that are considered most desirable. Attribute preferences and group preferences may thus be articulated prior to commencing the on-line procedure.

For example, suppose that a planner needs to make a selection of land parcels for a specific purpose and proposes to use the multiple-map overlay technique to facilitate the selection. For the sake of simplicity, imagine that the selection must be based on three factors—land use, land value, and population density. It may be further assumed that the planner considers land use twice as important as land value in the site selection process and four times as important as population density. This preference may be translated into weight factors of 4.00, 2.00, and 1.00 for land use, land value, and population density, respectively. If more maps were used in the multiple-map overlay process, a similar strategy would be applied to every other group of attributes (maps).

To determine appropriate weights, a panel of experts may have to rate each attribute independently, using an interval scale. The ratings of all experts may then be averaged to obtain the weight factors to be used in the overlay process. If serious discrepancies exist between the ratings of different experts, they may be resolved through negotiation and discussion.

A similar procedure must be conducted for comparative rating of attributes within each group. Here, assume that the planner has gathered expert ratings for each relevant attribute of each factor (on a rating scale of 0–15), reflecting the relative suitability of the particular attribute for the specific purpose. The most suitable attributes are rated with the highest values; lower values indicate attributes that are considered less suitable. The attributes and their ratings (by factor) are as follows:

Land Use
Vacant Land — 9
Residential Use — 8
Parks — 5
Industrial Land — 1

Land Value (per square yard)
Cost > $1000 — 1
$750 < Cost < 1000.00 — 5
$500 < Cost < 750.00 — 8
Cost < $500.00 — 15

Population Density (people per acre)
Population Density > 100 — 3
50 < Population Density < 100 — 5
10 < Population Density < 50 — 10

If more attributes were to be rated, the planner would use the same type of negotiated approach to derive comparative rating factors for each attribute.

The decision to use a rating scale of 0–15 in the above procedure has two justifications. First, it matches the numerical codes corresponding to the low-resolution graphics colors. As such, the comparative ratings of attributes can be graphically represented by converting the numerical value of the rating into the corresponding color code value. This would allow conversion of a thematic map into a comparative ratings map for that group of attributes.

Second, the comparative rating procedure is conducted on an interval scale; and when such a scale is used, the differences may be explained in comparative terms. For instance, one attribute may be regarded as x units better than another attribute, rather than as x times better than the other attribute. Multiplying interval variables by a weight factor preserves the relative differences. Thematic maps, on the other hand, are prepared on a nominal scale. Nominal scale variables cannot be subjected to mathematical manipulations without first being translated into interval scale variables.

Both the comparative rating process and the weight assignment process are highly subjective and reflect the bias of the user.[2] The finalized comparative ratings and weight factors become the user-supplied inputs most critical to the effective implementation of the multiple-map overlay utility.

On-Line Procedures

The relevant on-line procedures for operating the multiple-map overlay utility consist of five steps, discussed individually below.

Step 1. The selected attributes in the data map for each factor must be converted into priority maps. The numerical codes of displayed colors on this map reflect the comparative ratings assigned to each attribute. The process of rating attributes comparatively within a map makes use of the recode utility. The user recodes the original colors on the displayed data map, using colors corresponding to the comparative ratings established in the off-line procedure. This process is performed on every data map selected for use in the overlay procedure, and the recoded maps become the input source maps for the overlay process.

A section of a hypothetical land use data map is displayed in matrix A. R represents residential use; V represents vacant land; P represents parks; and I represents industrial use.

The recode utility may be invoked to convert the land use map into a comparative rating map that uses the colors corresponding to the rating codes. The comparative ratings for the portion of land use map shown in matrix A would appear as shown in matrix 1.

MATRIX A.

P	P	R	R
P	P	V	R
P	R	V	V
I	I	I	I

MATRIX 1.

5	5	8	8
5	5	9	8
5	8	9	9
1	1	1	1

Matrix B shows the land value classifications for each of the corresponding cells displayed in matrix A. The legend for this map is:

The Multiple-Map Overlay Utility

Cost > $1000.00 per sq.yd: A
Cost between $750.00 and 1000.00 per sq.yd: B
Cost between $500.00 and 750.00 per sq.yd: C
Cost < $500.00 per sq.yd: D

MATRIX B.

B	B	C	C
B	B	D	C
B	C	D	D
A	A	A	A

The recode utility may be invoked to convert the land value map into a comparative ratings map that uses colors corresponding to each of the rating codes. The comparative ratings for the portion of land value map shown in matrix B would appear as shown in matrix 2.

Population densities for the corresponding areas are shown in matrix C. The legend for the densities is as follows:

Population density > 100 ppa: X
Population density > 50 ppa and < 100 ppa: Y
Population density > 10 ppa and < 50 ppa: Z

The recode utility may be invoked to convert the population density map into a comparative ratings map that uses colors corresponding to each of the rating codes. The comparative ratings for the portion of population density map shown in matrix C would appear as shown in matrix 3.

MATRIX 2.

5	5	8	8
5	5	15	8
5	8	15	15
1	1	1	1

MATRIX C.

Y	Y	Z	Z
Y	Y	Z	Z
Y	Y	Z	Z
X	X	X	X

MATRIX 3.

5	5	10	10
5	5	10	10
5	5	10	10
3	3	3	3

Step 2. The process of assigning weights to a comparatively rated attribute map is performed during the overlay process. Priorities are assigned by entering (from the keyboard, in response to prompts from the screen) appropriate weight factors corresponding to attribute ratings on each data map.

Step 3. The algorithm for selecting the most suitable site is obtained in three steps. First, the rating of each cell of every comparatively rated map must be multiplied by the weight factor assigned to that map.

At this step, matrix 1 converts to matrix 4.

Similarly, matrix 2 converts to matrix 5.

Similarly, matrix 3 converts to matrix 6.

Step 4. The next step in the algorithm involves a simple matrix addition, whereby the numbers in each cell of matrix 4, matrix 5, and matrix 6 are added together to produce matrix 7.

Step 5. The final step in the algorithm is to normalize the scores in each cell. This is performed by dividing the scores in each cell by the sum of the weights used. In this instance, the divisor is 7—the sum of weights 4, 2, and 1. The results, rounded off to the closest integer value, are displayed in matrix 8.

These values serve as the numerical codes for generating a low-resolution graphics display.

MATRIX 4.

20	20	32	32
20	20	36	32
20	32	36	36
4	4	4	4

MATRIX 5.

10	10	16	16
10	10	30	16
10	16	30	30
2	2	2	2

MATRIX 6.

5	5	10	10
5	5	10	10
5	5	10	10
3	3	3	3

MATRIX 7.

35	35	58	58
35	35	76	58
35	53	76	76
9	9	9	9

MATRIX 8.

5	5	8	8
5	5	11	8
5	8	11	11
1	1	1	1

INTERPRETATION OF GRAPHICS DISPLAYS

Matrix 8 displays the composite scores for each cell, based upon the weights and ratings used to reflect comparative suitability. In this case, the most suitable land has a composite score of 11 and the least suitable land has a composite score of 1. In the graphics conversion, colors having higher numerical code equivalents (such as white, aqua, and yellow) depict locations that are most likely to satisfy the user for the proposed need; conversely, colors having lower numerical code equivalents (such as black, dark red, and dark blue) depict locations that are least likely to satisfy the expectations of the user. Intermediate values indicate diminishing levels of suitability for the designated purpose.

The display reflects a solution based on the assumptions and value judgments implicit in the overlay process. Alternative modifications and compromises in ratings and in the assignment of weights to thematic maps and attributes may be made. A series of map overlay products that reflect all such modifications and compromises can be produced in this way, providing a wide range of choices for comparison and evaluation before final site selection is made.

The following subroutine enables a user to activate the multiple-map overlay utility:

```
10000 REM *** SUBROUTINE OVERLAY ***
10010 REM THIS SUBROUTINE PERFORMS MULTIPLE
      OVERLAY
10020 PRINT D$;"PR#0": PRINT D$;"IN#0"
10030 INPUT "ENTER NUMBER OF THE MAPS TO BE
      OVERLAID"; N$
10040 K = VAL (N$): HOME
10045 NF = 0 : REM RESET NORMALIZATION FACTOR
10050 IF K = 0 THEN 10380
10055 FOR I = R TO R4%: REM CLEAR MATRIX D%
10060     FOR J = R TO R4%
10062         D%(I,J) = 0
10064     NEXT
10068 NEXT
10070 FOR NK = 1 TO K
10080     HOME
10090     INPUT "ENTER NAME OF THE MAP ?";NN$
10095     GOSUB 1300: PRINT
10100     INPUT "ENTER MAP'S WEIGHT FACTOR
          ?";A$
10110     N = VAL (A$)
10120     NF = NF + N
10125     NN$ = "MAP." + NN$
10130     PRINT: PRINT D$"BLOAD";NN$;
          ",A1024,D";DR$
10140     HOME: PRINT "<< PLEASE WAIT >>"
10160     FOR I = R TO R4% : REM PERFORM
          MATRIX ADDITION
10170         FOR J = R TO R4%
10180             E%(I,J) = SCRN (I,J)
10190             D%(I,J) = .5 + D%(I,J) +
                  E%(I,J) * N
10200         NEXT
10210     NEXT
10220 NEXT
10225 IF NF = 0 THEN 10380: REM ILLEGAL INPUT
10230 FOR I = R TO R4% : REM PERFORM
      NORMALIZATION
10240     FOR J = R TO R4%
10250         D%(I,J) = D%(I,J) / NF
10260     NEXT
```

```
10270 NEXT
10280 INPUT "ENTER THRESHOLD VALUE FOR DISPLAY
      (0-15) ?";A$: N = VAL (A$)
10290 FOR J = R TO R4%
10300     FOR I = R TO R4%
10305         COLOR = 0: PLOT I,J
10310         IF D%(I,J) >= N THEN
              COLOR    = D%(I,J): PLOT I,J
10320     NEXT
10330 NEXT
10340 HOME: PRINT "DO YOU WISH TO SAVE THIS
      OVERLAY (Y/N) ?": GOSUB 1400: PRINT
10350 IF A$ = "Y" THEN PRINT: GOSUB 23000:
      PRINT D$;"PR#0":  PRINT D$;"IN#0": REM
      SAVE THE DISPLAY
10360 HOME: PRINT "DO YOU WISH TO CONTINUE
      (Y/N) ?": GOSUB 1400: PRINT
10370 IF A$ = "Y" THEN 10280
10380 PRINT: HOME:GOSUB 970: REM REDRAW THE
      MAP
10390 PRINT D$;"PR#";SL: PRINT "N": PRINT
      D$;"IN#";SL
10400 RETURN
```

Line 10030 inquires about the number of comparatively rated maps that are to be submitted to the overlay process. A null entry to this query terminates the process. The loop of lines 10055–10068 clears the data matrix D% to be used for matrix addition. The loop of lines 10160–10220 performs the matrix addition. The loop of lines 10230–10270 performs the normalization. Line 10280 seeks a threshold value (from 0 to 15) to be used as a filter for the final display. Only cells bearing a normalized score equal to or greater than the threshold value are filtered through for display. The filtering procedure is performed by the loop of lines 10290–10330. The product map may be saved, and the user may either continue to review product maps with new threshold filters or terminate the overlay procedure.

TYPICAL APPLICATION OF THE MULTIPLE-MAP OVERLAY UTILITY IN REGIONAL PLANNING

The authors first tested the multiple-map overlay utility in a regional planning application with students from the Department of Community and Regional Planning at the University of Nebraska. The objective of the proposed application was to delineate a "formal region" representing the Sand Hills of Nebraska in order to investigate the nature, characteristics, problems, and developmental prospects of cattle ranching in the area.

The Sand Hills of Nebraska comprise an area of more than 19,000 square miles that spans much of the state. The area constitutes one of the largest unbroken expanses of grassland in the United States. It is sparsely populated, and its agricultural economy places major emphasis on cattle ranching.

The Sand Hills area is subject to unique circumstances relating to surface water. The region receives very little rainfall; at the same time, the prevailing, very porous sandy soil has extremely poor water retention characteristics. These adverse atmospheric moisture and soil conditions render the area unfit for any form of committed food-crop farming.

In spite of these seemingly debilitating features, the groundwater endowments of the area present an entirely different scenario. The Sand Hills area is located on the largest and richest underground aquifer in the country—the Ogallala aquifer—which extends from North Dakota to Texas through Nebraska. The depth at which groundwater lies below the surface in the Sand Hills varies from one location (where water may be available at the surface itself) to another (where the water level may be several hundred feet below the surface).

The planning team was required to delineate a region that typified "cattle ranching country" within the Sand Hills area. In order to do so, six specific descriptors were designated:

1. The sand hills classification of topography was identified as the best topographical type for cattle range country in the Sand Hills region.
2. The availability of large quantities of groundwater from the Ogallala aquifer was designated as the key descriptor within the groundwater availability category. Groundwater represents the predominant water source in the Sand Hills region.
3. The rangeland classification—characterized by natural grasses and including improved pasture, forage crops, or special public use cover—was identified as the appropriate descriptor of cattle country within the general land use category.[3]
4. Sand, sand-silt, and sandstone were designated as the descriptors of the geological formations characteristic of cattle range country.
5. The sand hills designation chosen by the Conservation and Survey Division of the University of Nebraska was accepted as the most appropriate descriptor within the underground water area category.

6. A value of approximately 280 salt particles per million was identified as the threshold value for typical groundwater quality under irrigated and chemically fertilized rangeland.

Maps displaying these characteristics were secured from the Conservation and Survey Division of the University of Nebraska to serve as input source maps for digitization. The following maps were generated using the map creation and editing utilities described in this book: topographic regions map (see figures C-2 and 9-1); groundwater availability map (see figure 9-2); land use classification map (see figure 9-3); map of soil types by geological formation (see figure 9-4); underground water classification map (see figure 9-5); and groundwater quality map (see figure 9-6).

The maps were saved using the save utility and collectively became the graphic data base for the region delineation exercise.

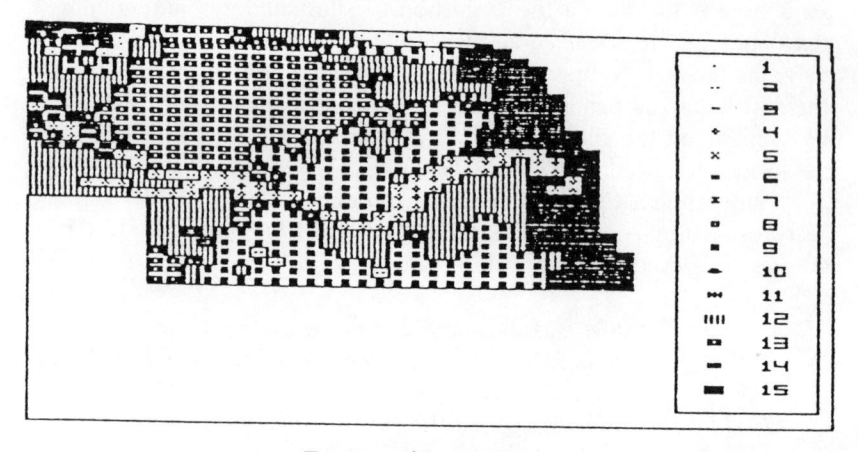

Topographic Regions

LEGEND

#	Key	Freq.
1	Bluffs & Escarpments	17
2	Large Reservoirs	5
4	Valleys	62
9	Dissected Plains	190
12	Plains	201
13	Sand Hills	211
14	Valley-side Slopes	26
15	Rolling Hills	113

Figure 9-1. Hard copy of topographic regions.

The Multiple-Map Overlay Utility

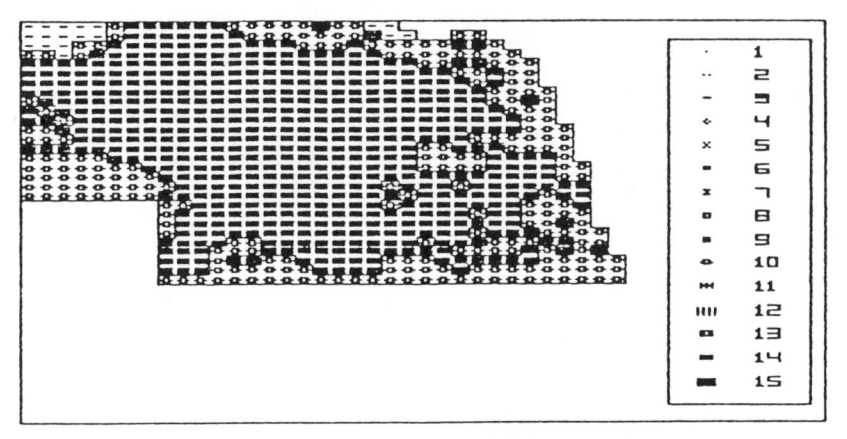

Groundwater Availability

LEGEND

#	Key	Freq.
3	Small Quantities	18
10	Medium Quantities	237
14	Large Quantities	557

Figure 9-2. Hard copy of groundwater availability.

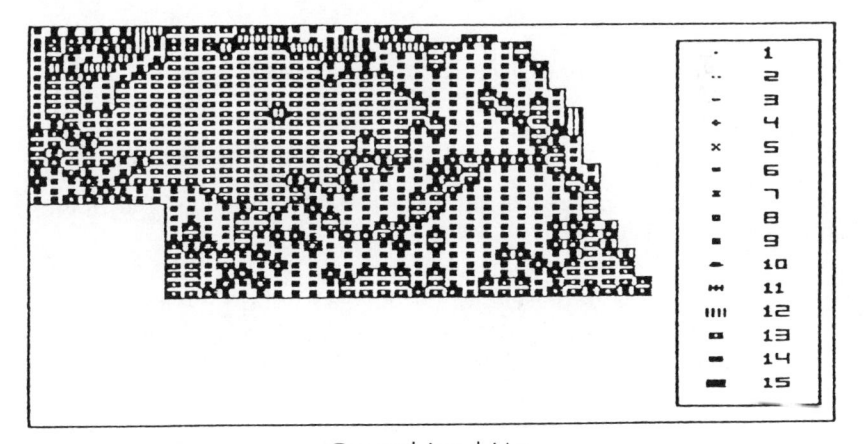

General Land Use

LEGEND

#	Key	Freq.
9	Agriculture	386
12	Forest	23
13	Rangeland	437

Figure 9-3. Hard copy of land use classification.

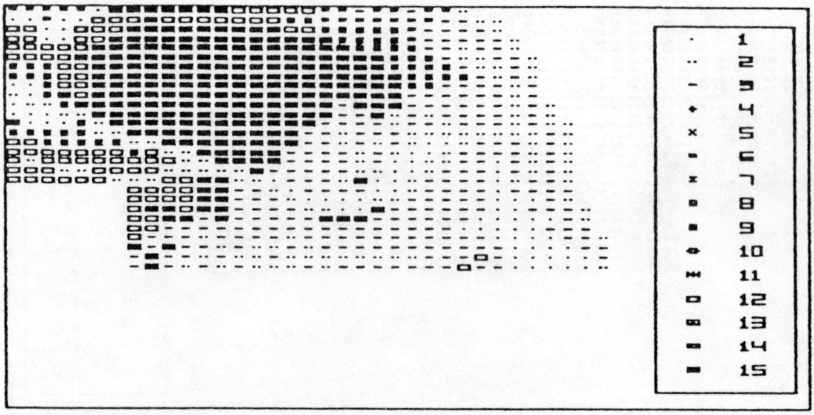

116　CARTOGRAPHY AND SITE ANALYSIS WITH MICROCOMPUTERS

Soil Parent Materials

#	LEGEND Key
1	Loess (Drift & Alluvium)
2	Alluvium
3	Loess
6	Shale
9	Sand & Silt
12	Sandstone
15	Sand

Figure 9-4. Hard copy of soil type by geological formation.

Assigning Weights in the Regional Planning Example

Assigning weights that reflect the priorities of each group of attributes was made as follows:

Weights of 15—representing the highest priority—were assigned to topographic region (map 1), land use (map 3), geological formation (map 4), and underground water classification (map 5). Secondary importance—characterized numerically by a weight of 10—was given to groundwater availability (map 2). A weight of 7—representing the lowest importance—was assigned to groundwater quality (map 6). These weights were determined through a negotiated process involving the students, who served as "the experts"; various alternatives were examined before the above weighting scheme was accepted.

Selected attributes were chosen from each thematic map, and each attribute was assigned a weight multiplier to reflect its importance relative to other attributes represented in the map. The sand hills classification

The Multiple-Map Overlay Utility

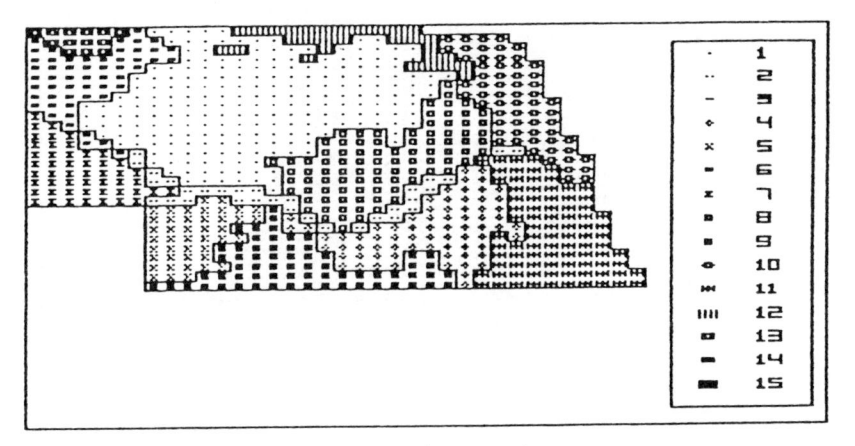

Underground Water Area

#	Key	Freq.
1	Sand Hills	228
2	Platte R.V. Region	32
4	S.C. Plains Region	73
5	S.W. Tableland Region	43
6	N. Panhandle T.R.	54
7	Panhandle T.R.	50

Figure 9-5. Hard copy of underground water classification area.

was assigned a weight of 15 in the topographic regions map; the sand hills classification and rangeland classification were both assigned weights of 15 in the underground water and land use maps, respectively. Sand was assigned a weight of 15, sand-silt a weight of 10, and sandstone a weight of 5 in the soil types map. Finally, the availability of large quantities of groundwater and a salt content of less than 280 salt particles per million (sppm) were each assigned weights of 15 in the groundwater availability map and groundwater quality map, respectively.

The weights assigned to the various relevant attributes were used to convert each thematic map into a prioritized map describing a particular group of attributes. Six priority maps were thus created, using the recode utility. Attributes not assigned a priority were recoded in black, symbolizing no priority.

The multiple-map overlay utility was then invoked to determine the location of a coherent and contiguous area that best represented "cattle country" in the Sand Hills region. The appropriate weights were given to the computer in response to program prompts for each group of

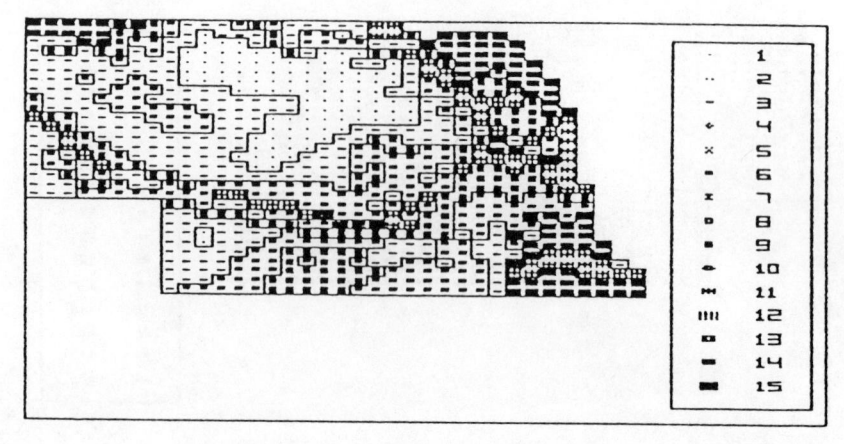

Groundwater Quality

LEGEND

#	Key	Freq.
1	0–170 SPPM	105
3	170–280 SPPM	264
6	280–400 SPPM	189
9	400–540 SPPM	127
11	540–700 SPPM	69
14	700 SPPM and Over	91

Figure 9-6. Hard copy of groundwater quality.

attributes (maps) as each was retrieved for conducting the overlay procedure.

A sliding interval scale for ranking aggregations of desired attributes and groups was then established. The units ranged from 0 to 15, in increments of 1. A value of 15 would signal that all priorities were fulfilled at that cell location, whereas a value of 0 would signal that no priority was fulfilled. Intermediate values would suggest mixed levels of priority fulfillment.

Successive threshold values of 15 through 12 were designated as cutoff points for priority fulfillment. Any cell that displayed a priority fulfillment value of less than 12 was considered to have inadequately fulfilled the priority for purposes of the delineation process and was to be discarded. All cells that equaled or surpassed the threshold value were to be displayed.

Figure 9-7 displays the cells whose priority fulfillment scores equaled a threshold value of 15; figure 9-8 displays those whose scores equaled

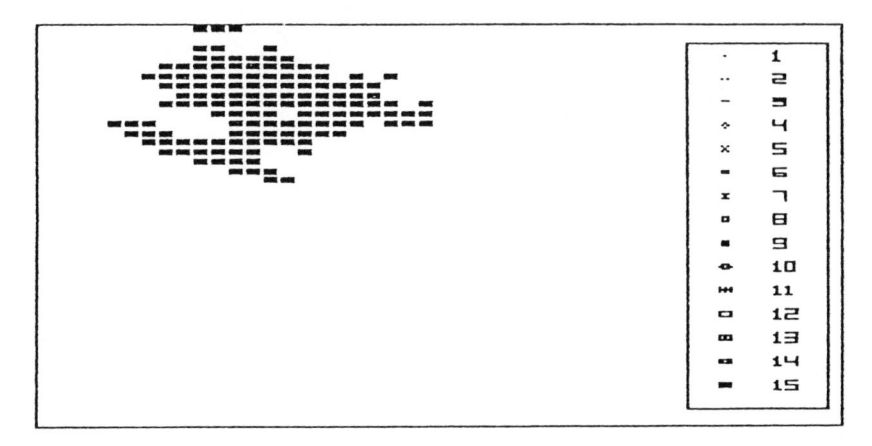

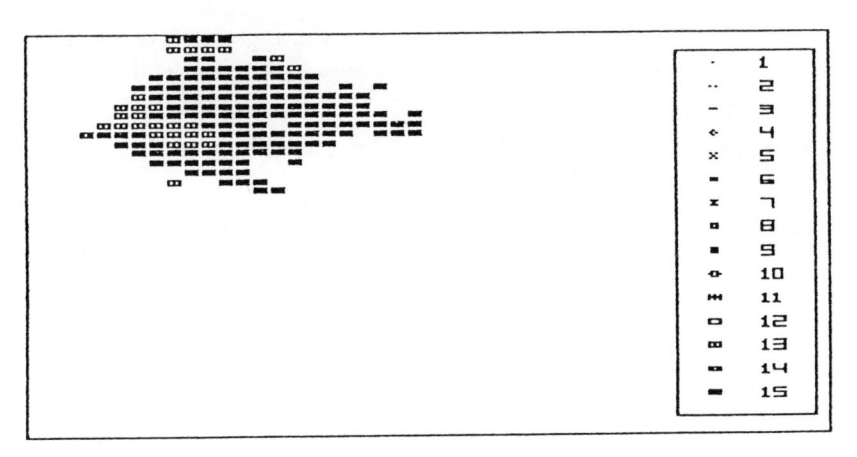

Figure 9-7. Hard copy of region with a threshold value of 15.

or exceeded a threshold value of 13; figure 9-9 displays those whose scores equaled or exceeded a threshold value of 12.

Next, a contiguity and coherence adjustment was made on the overlay product map that used a threshold value of 12. The resulting map, showing the adjusted delineation of the formal cattle ranching region of the Sand Hills area, is displayed in figure C-3 of the color insert.

The map displaying the delineated region of the Sand Hills seemed visually sterile and noninformative for want of such reference features

Figure 9-8. Hard copy of region with a threshold value of 13.

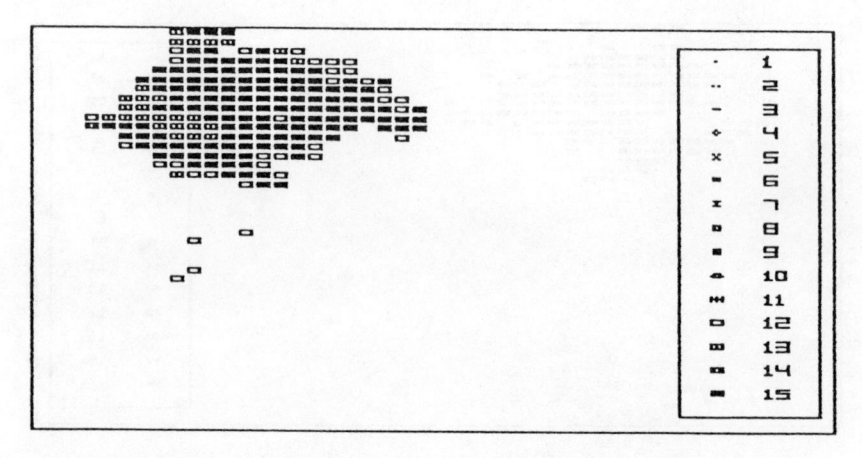

Figure 9-9. Hard copy of region with a threshold value of 12.

as jurisdictional boundaries. In order to overcome this visual shortcoming, two additional maps were generated and saved by means of the map create, edit, and save utilities. Figure C-4 of the color insert depicts the jurisdictions of natural resource districts in the northwestern part of Nebraska. Figure C-5 of the color insert depicts county boundaries in the state.

The opaque overlay utility was invoked to perform opaque overlays of the map showing the delineated region of cattle country on each of the two maps showing jurisdictional divisions. Figure C-6 of the color insert displays the opaque overlay of the delineated region on the natural resource districts map of northwestern Nebraska. Figure C-7 of the color insert displays the opaque overlay of the delineated region on the county map of the state.

Assessment of the Results of the Regional Planning Example

The process of map generation, editing, multiple-map overlaying, and opaque overlaying performed by various special utilities considerably reduced the time required to perform the delineation exercise. Each utility was called upon to perform specific components of an esoteric task; considerable judgment was exercised in choosing the appropriate utility and applying it judiciously to perform the necessary tasks.

The result of the process was a visually comprehensible, technically justifiable, and logically defensible delineation of a formal region—a delineation that negotiated all the complexities of prioritization and selection. Even though the delineation procedure itself was only a demonstration exercise, it possessed sufficient complexity to benefit from a

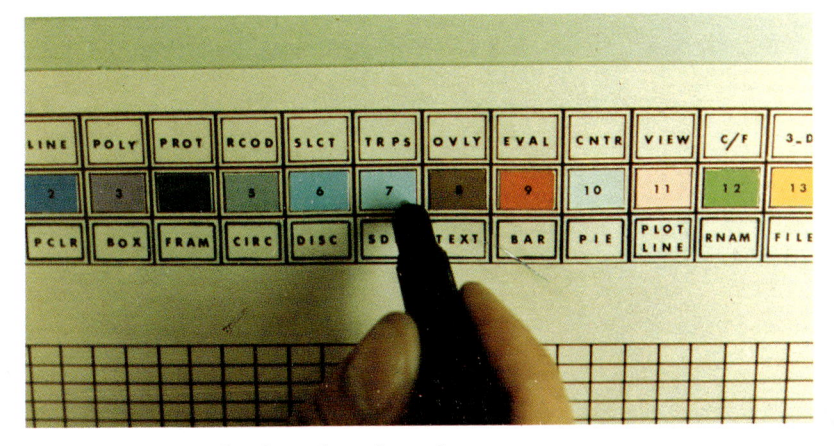

C-1. Accessing a color from the color palette.

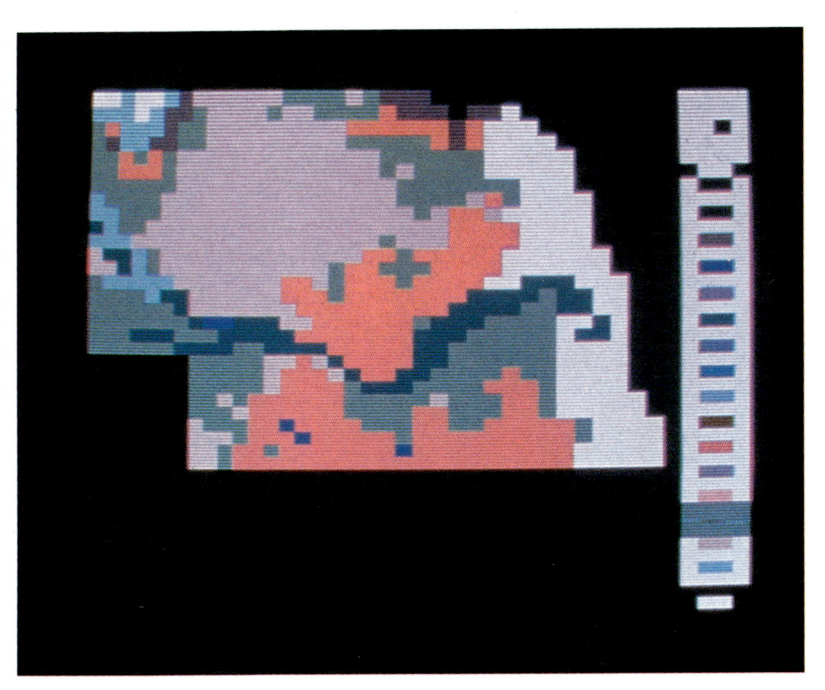

C-2. Typical map output after use of the map creation and map editing features. LEGEND: magenta, bluffs & escarpments; dark blue, large reservoirs; dark green, valleys; orange, dissected plains; green, plains; yellow, sand hills; aqua, valley-side slopes; white, rolling hills.

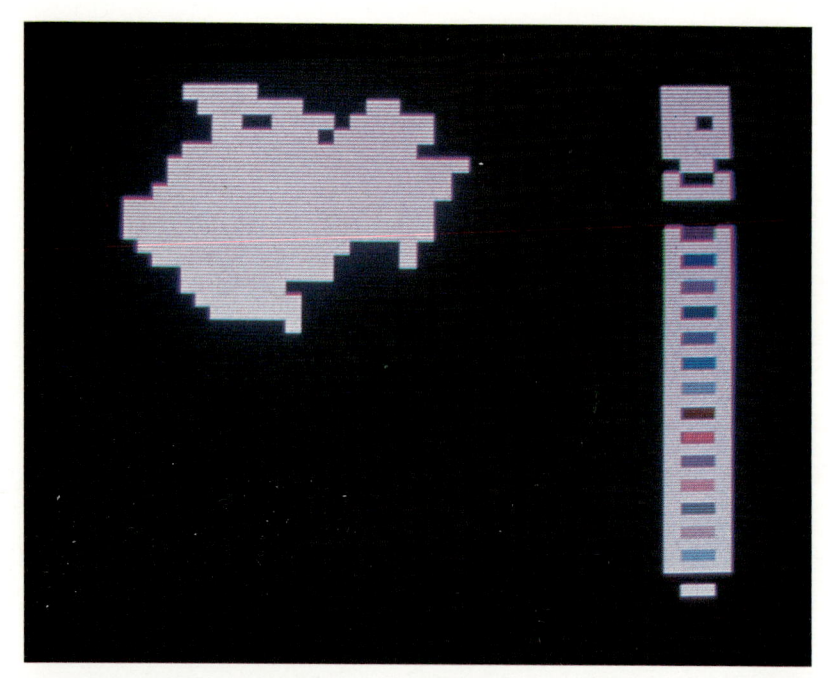

C-3. *Delineated sand hills region.*

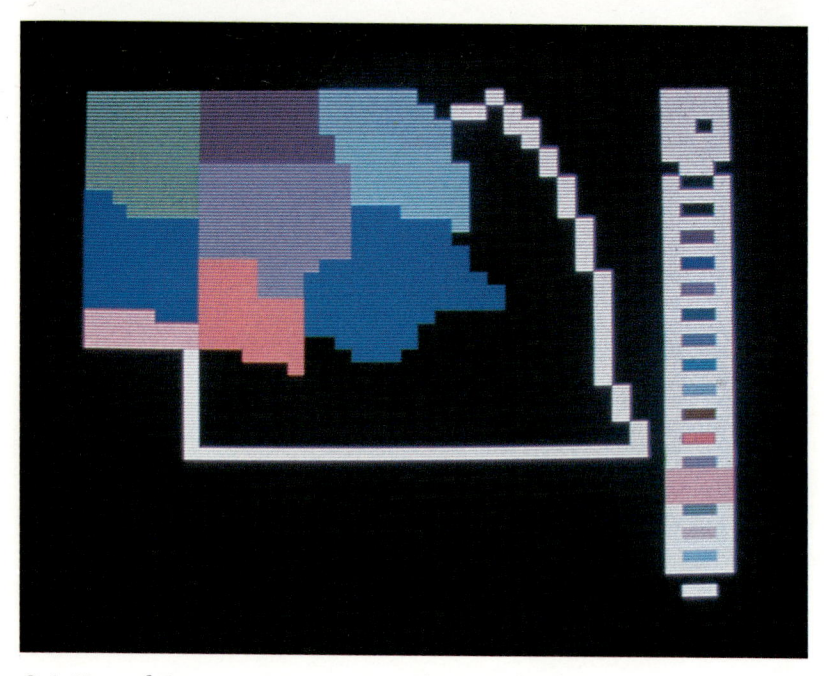

C-4. *Natural Resource Districts in northwest Nebraska.*

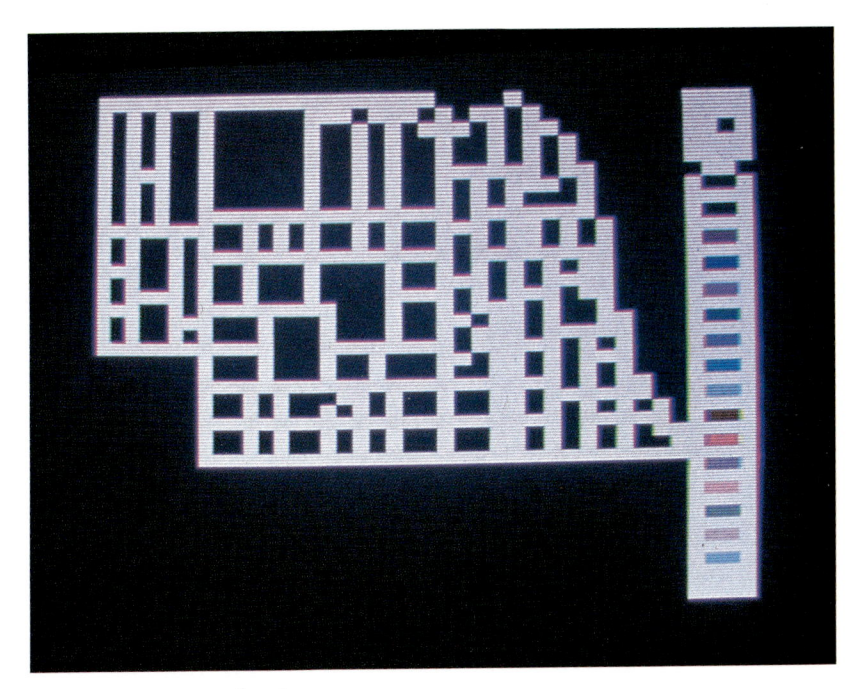

C-5. Counties in Nebraska.

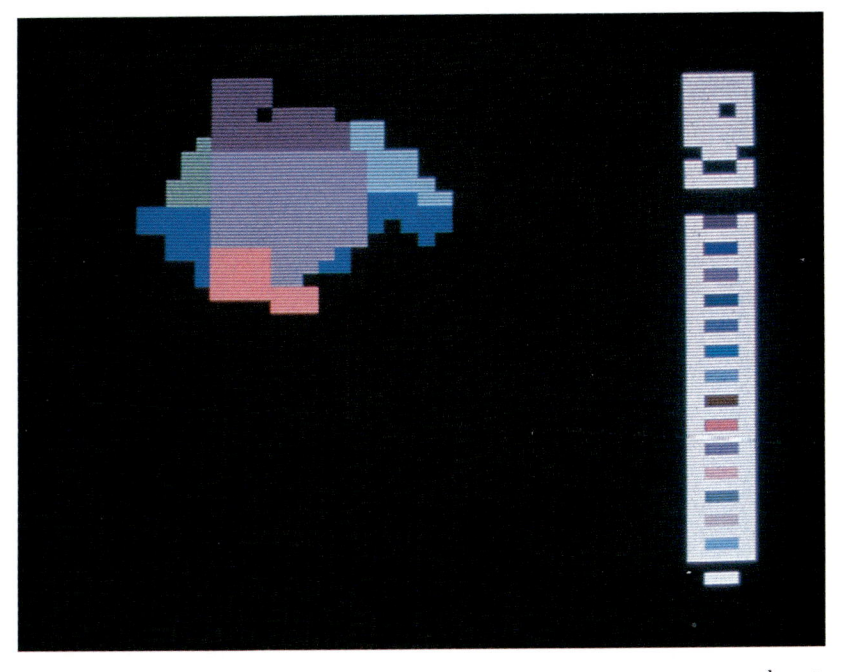

C-6. Delineated region showing included Natural Resource Districts in northwest Nebraska.

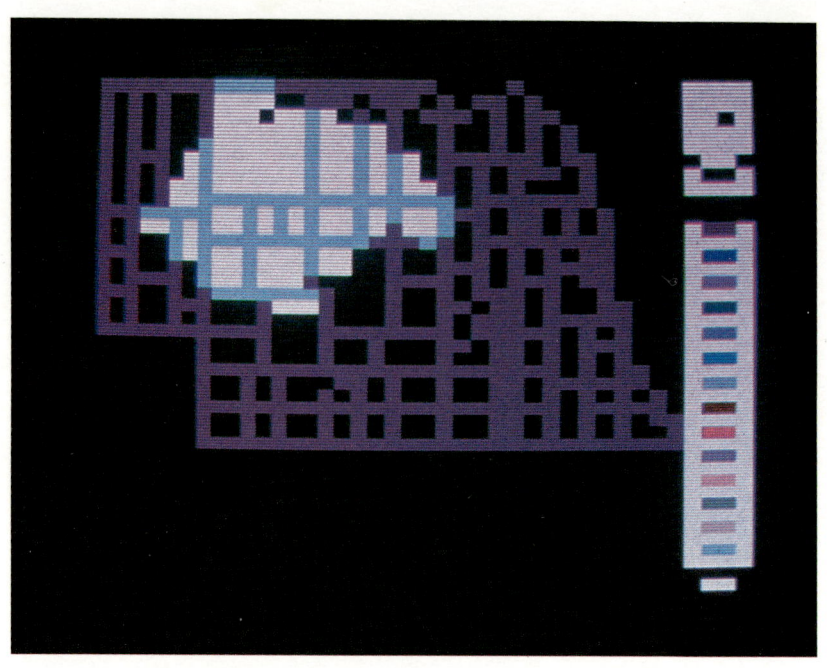

C-7. *Delineated region showing included counties.*
LEGEND: *purple, county lines outside delineated boundary; aqua, county lines within delineated boundary; white, delineated sand hills region.*

powerful, rapidly generated computer-aided solution. The demonstration suggests that regional planning is one of several areas in planning and design that could make advantageous use of the utilities described in this book, with considerable productivity gains in problem-solving attainable in the process.

NOTES

1. Ian L. McHarg, *Design with Nature* (Garden City, N.Y.: Doubleday, 1971).
2. It has frequently been argued that the overlay process involves a rating and weighting process that is unscientifically subjective and biased. While the authors acknowledge that site selection is and always will be subjective and value-oriented and believe that scientific objectivity plays a crucial part in the final decision-making process, they would point out that value judgments usually tend to supersede scientific objectivity in any case. Some value judgments are more defensible than others, however, and heightened defensibility renders superior judgments.

The process suggested by the authors involves debate, negotiation, and compromise among experts to derive weighting factors. Alternative procedures are also available.

A good article on the subject of map overlaying is Ardeshir Anjomani, "The Overlaying Map Technique: Problems and Suggested Solutions," *Journal of Planning Education and Research* 4(2) (December 1984): 111–19.
3. Bluestem grasses (*Andropogon*), sandreed grass (*Calamovilfa*), needle grass (*Stipa*), and yucca (*Yucca*) are the principal natural grasses in the Sand Hills area.

TEN

A Utility for Evaluating Alternative Sites According to Specific Criteria

Computer cartography and site analysis applications have at least two thrusts of potential aid to planners and designers in the decision-making process. First they enable users to create a data base of thematic maps and legends relating to the targets of study. Second, they facilitate quick and convenient storage, retrieval, and updating of thematic maps and legends.

Analytical applications for site selection may be conducted by subjecting selected thematic maps from the data base to scans and overlays. The products of these analytical procedures establish a new data base of site and location alternatives capable of satisfying diverse priorities, preferences, and final objectives.

The purpose of evaluation utilities is to assess various alternative sites based on selected criteria. The user may then rank the alternatives based on these comparisons.

THE EVALUATION UTILITY

The evaluation utility is a vehicle for evaluating, comparing, and ranking a number of alternative sites so that the user can make a justifiable choice among them. The evaluation utility uses only comparatively rated maps of natural or altered sites.[1] The principle of comparative rating used by the evaluation utility is similar to that used by the overlay utility.

Comparatively rated maps may be derived from any of three sources: maps that have been generated as data maps by the user through the map creation and editing utilities, and then recoded by means of the recode utility to reflect comparatively rated suitability; maps that have been subjected to intentional attribute modification by the user through the editing utilities, and then recoded by means of the recode utility to reflect comparatively rated suitability; and maps that have evolved as results of the alternative generation process.

This last map source comprehends maps derived through the logical overlay process; the scanning process, followed by the comparative rating process involving the recode utility; the scanning process in tandem with the opaque overlay process, followed by the comparative rating process involving the recode utility; or some combination of the above processes.

The utility has the capacity to evaluate and compare up to twenty different alternatives at a time.

THE EVALUATION ALGORITHM

The objective of the evaluation utility is to derive absolute and relative indices of comparison for each alternative, based on certain chosen criteria for evaluation.

The utility uses the numerical codes corresponding to particular colors as indicators for the scoring process. Numerical code values of all cells are summed to obtain a CSCORE. The SCORE is derived from the CSCORE using the formula:

$$SCORE = CSCORE \div (NUMBER\ OF\ CELLS)$$

The SCORE is an absolute index of aggregate cell suitability. Only cells not colored black (that is, cells having numerical value equivalents greater than zero) are included in the above calculation. The higher the numerical value of the SCORE, the better the alternative.

A secondary indicator—the standard deviation of the SCOREs—is also derived by the utility. This indicator provides the user with a relative index of the homogeneity that prevails among the different cells in each of the evaluated alternatives.

The following subroutine enables a user to activate the evaluation utility:

```
11000 REM *** SUBROUTINE EVALUATION ***
11010 REM THIS SUBROUTINE EVALUATES
      ALTERNATIVES
11020 PRINT D$;"PR#0": PRINT D$;"IN#0"
11030 INPUT  "ENTER  NUMBER OF ALTERNATIVES TO
      BE EVALUATED ? "; N$
11040 K = VAL (N$): HOME
11050 IF K = 0 THEN 11490
11055 IF  K  >  20  THEN  11030:    REM NOT
      MORE THAN 20 ALTERNATIVES MAY BE
      EVALUATED
11060 FOR NK = 1 TO K
11070     HOME
11080     PRINT "ENTER NAME OF THE MAP # ";NK;
11090     INPUT A$(NK)
11100     Q%(NK) = 0: REM INITIALIZE SCORE
          VARIABLE
11102     P%(NK) = 0: REM INITIALIZE TOTAL
          SCORE (CSCORE)
11104     S(NK) = 0: REM INITIALIZE STANDARD
          DEVIATION
11110 NEXT
11120 GOSUB 1300: PRINT
11122 P% = 0:REM INIT. AVERAGE SCORE
11124 Q% = 0:REM INIT. AVERAGE CSCORE
11130 FOR NK = 1 TO K
11140     NN$ = "MAP." + A$(NK)
11150     PRINT D$"BLOAD";NN$;",A1024,D";DR$
11160     HOME: PRINT "<< PLEASE WAIT >>"
11170     C = 0
11180     FOR J = R TO R4%
11190         FOR I = R TO R4%
11200             E%(I,J) = SCRN (I,J)
11210             IF E%(I,J) <> 0 THEN
                  C = C + 1:
                  P%(NK) = P%(NK) + E%(I,J):
                  S(NK) = S(NK) + (E%(I,J)^2
11220         NEXT
11230     NEXT
11240     IF C <= 1 THEN 11490: REM ERROR
11250     N = ABS ((C * S(NK) - P%(NK) ^
          2) / (C * ( C - 1 ))):
```

```
                REM  CALCULATE  STANDARD  DEVIATION
11260           S(NK) = INT(100 * SQR(N)) / 100: REM
                ROUND OFF TO TWO DECIMAL POINTS
11270           Q%(NK) = P%(NK)
11280           P%(NK) = .5 + P%(NK) / C:  REM ROUND
                OFF
11290           P% = P% + P%(NK)
11300           Q% = Q% + Q%(NK)
11310 NEXT
11320 P% = .5 + P% / K
11330 Q% = .5 + Q% / K
11340 TEXT: HOME
11350 INVERSE:  HTAB  8:  PRINT  "
      ALTERNATIVE DESIRABILITY "
11360 PRINT"ALTERNATIVE          SCORE     ST.D
      CSCORE"
11365 NORMAL
11370 FOR NK = 1 TO K
11380     PRINT A$(NK);
11390     HTAB 20: PRINT P%(NK);
11400     HTAB 30: PRINT S(NK);
11410     HTAB 35: PRINT Q%(NK)
11420 NEXT
11430 PRINT "AVERAGE";
11440 HTAB 20: PRINT P%;
11450 HTAB 35: PRINT Q%
11460 VTAB 24: HTAB 10: PRINT "<< PRESS ANY
      KEY >>";
11470 GET A$: PRINT
11480 GR: XM = DC%: GOSUB 21000: GOSUB 970
11490 PRINT: HOME
11500 PRINT D$;"PR#";SL: PRINT "N": PRINT
      D$;"IN#";SL
11510 RETURN
```

The program is self-explanatory.

A TYPICAL APPLICATION OF THE EVALUATION UTILITY

A demonstration exercise of the evaluation utility was conducted by the students of the Department of Architecture at the University of Nebraska

126 CARTOGRAPHY AND SITE ANALYSIS WITH MICROCOMPUTERS

at Lincoln, in the realm of urban suitability studies. The principal objective of the study was to determine which areas in part of the Gretna quadrangle of the Omaha–Council Bluffs region were suitable for urban use. Three major tasks had to be undertaken consistent with the prescriptions for the conduct of environmental resource analysis: inventory; resource factor analysis; and suitability assessment.[2]

The study area was defined by the following boundaries: U.S. Highway 6 on the west; Nebraska State Highway 370 on the south; 156th Street on the east; and West Center Road on the north.

The area involved covers approximately 25 square miles. In order to make comparisons, the study area was divided longitudinally (north-south) into three contiguous windows, each window representing a component of a three-part mosaic. Each window was treated as an input source map for digitization and as an area capable of delivering an alternative site for urban use.

Six groups of attributes were selected as criteria for determining urban suitability: soil type; floodplain; slope; vegetation; surface water; and noise. Input source maps were collected from the Omaha–Council Bluffs Metropolitan Area Planning Agency (MAPA) for each of the six criteria, and data maps were generated using the map creation and editing utilities described previously.

Each map displayed variations in each of the six characteristics graphically represented. The *soil* characteristic was categorized as most suitable, marginally suitable, and least suitable. Distinctions among suitability levels were made based on "soil limitation ratings" made in a special

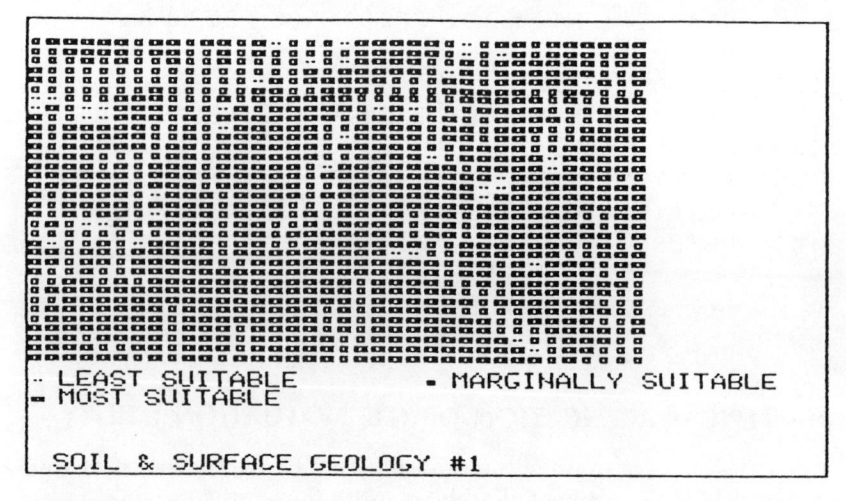

Figure 10-1. Hard copy of classification of soils—Lower Gretna quadrangle.

study by MAPA on the basis of shallow excavations for local roads, streets, and dwellings. The following "urban use averages" were mathematically derived (and the associated urban suitabilities of soils prescribed) by MAPA: values between 1.00 and 1.29 were considered most suitable; values between 1.30 and 1.50 were considered marginally suitable; values exceeding 1.5 were considered least suitable.[3]

Figures 10-1, 10-2, and 10-3 are hard copies of the soil maps generated for each study area.

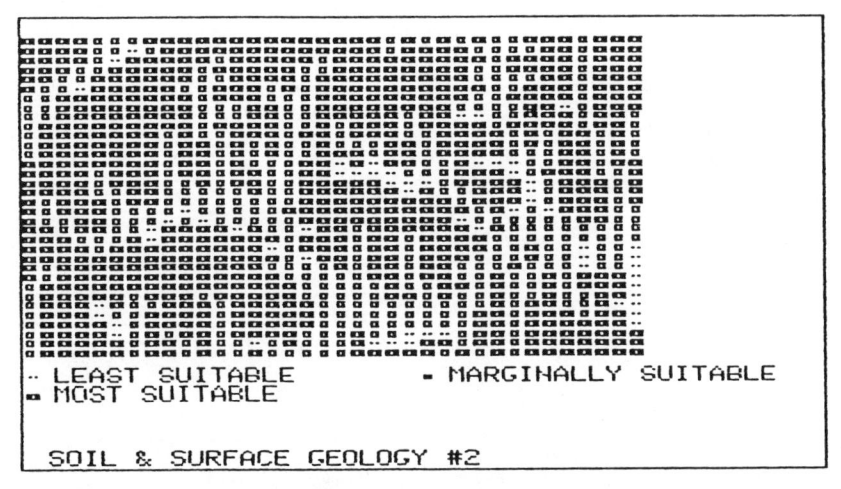

Figure 10-2. Hard copy of classification of soils—Middle Gretna quadrangle.

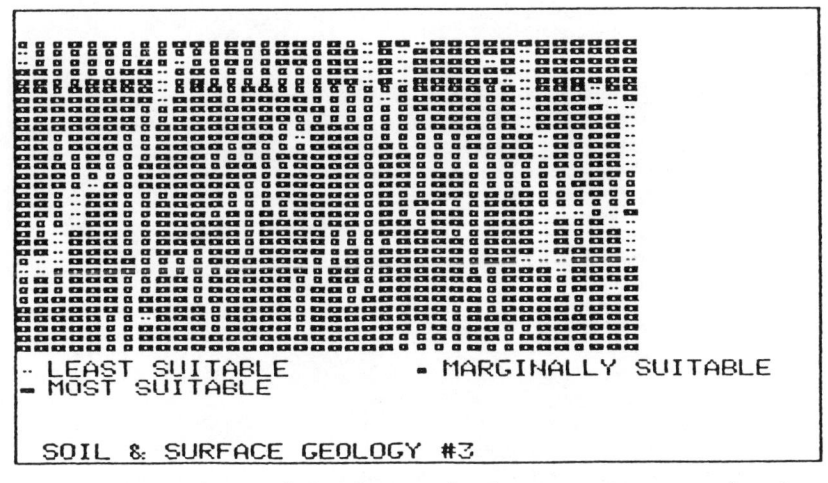

Figure 10-3. Hard copy of classification of soils—Upper Gretna quadrangle.

Three categories of *floodplains* were identified: areas not subject to flooding, areas likely to be flooded, and areas inundated. Areas not subject to flooding were rated as most suitable.[4]

Figures 10-4, 10-5, and 10-6 illustrate the floodplain characteristics for each study area.

The third factor taken into account in determining urban suitability was *slope*. Here again, three categories were identified: slope less than

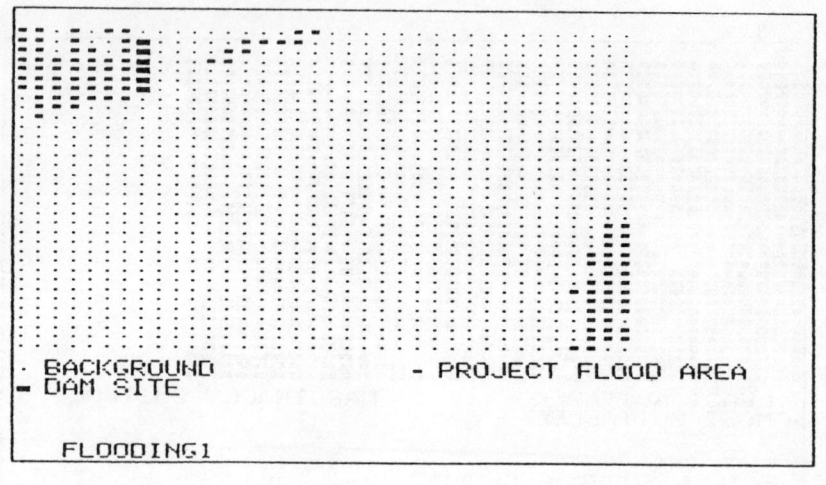

Figure 10-4. Hard copy of floodplain characteristics—Lower Gretna quadrangle.

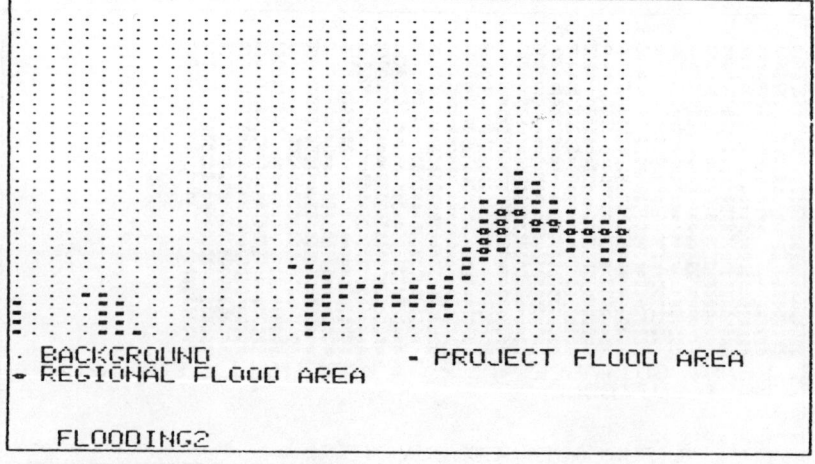

Figure 10-5. Hard copy of floodplain characteristics—Middle Gretna quadrangle.

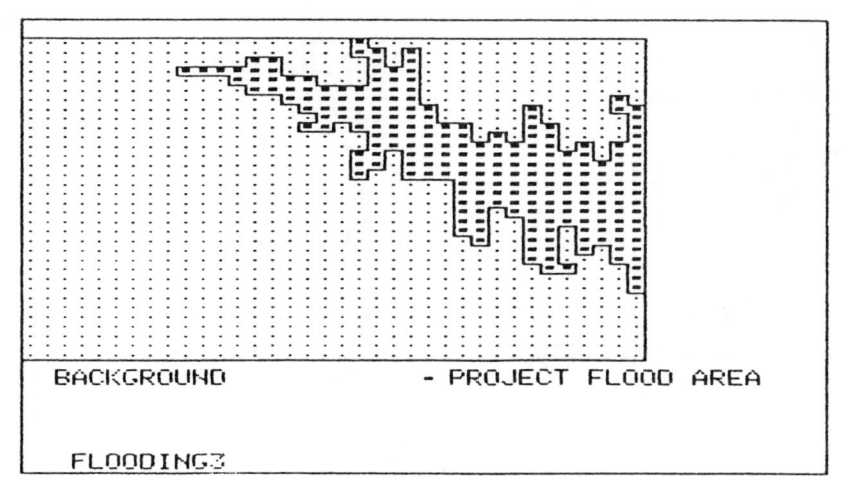

Figure 10-6. Hard copy of floodplain characteristics—Upper Gretna quadrangle.

7 percent, slope between 7 and 17 percent, and slope greater than 17 percent.

Although all three slopes did allow urban use, areas with minimal slope were considered most appropriate, while those with significant slope were considered least suitable.[5]

Figures 10-7, 10-8, and 10-9 illustrate the slope characteristics for each study area.

The fourth factor considered was *cover vegetation*. Three categories of land were identified: little or no tree cover, moderate tree cover, and heavy tree cover.

It was decided that areas with the least tree cover were most appropriate for urban use, while areas with considerable tree cover were inappropriate.[6]

Figures 10-10, 10-11, and 10-12 illustrate the classification of cover vegetation in each study area.

The fifth factor considered was *surface water*. Four categories of land were identified and classified within this characteristic: dry land; intermittent streams; perennial streams; and lakes, reservoirs, ponds, swamps, and marshes.

It was decided that dry lands were most appropriate for urban use, while all other areas became progressively less appropriate as surface wetness increased.[7]

Figures 10-13, 10-14, and 10-15 illustrate the distribution of surface water in each study area.

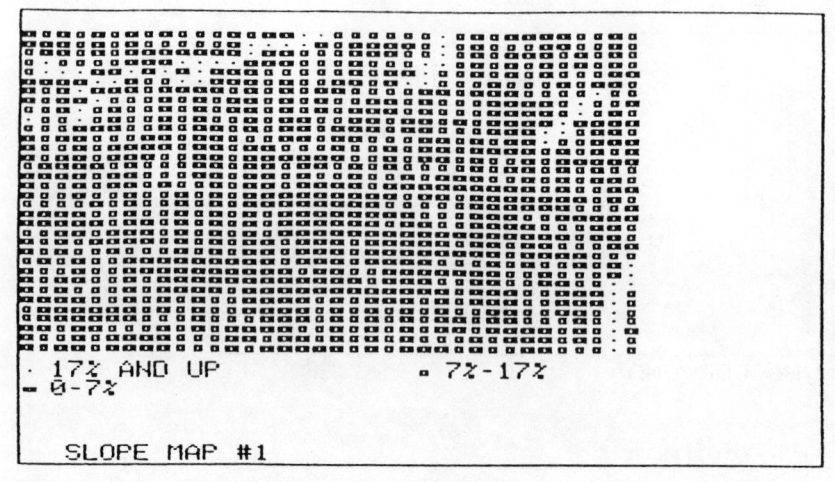

130 CARTOGRAPHY AND SITE ANALYSIS WITH MICROCOMPUTERS

Figure 10-7. Hard copy of slope characteristics—Lower Gretna quadrangle.

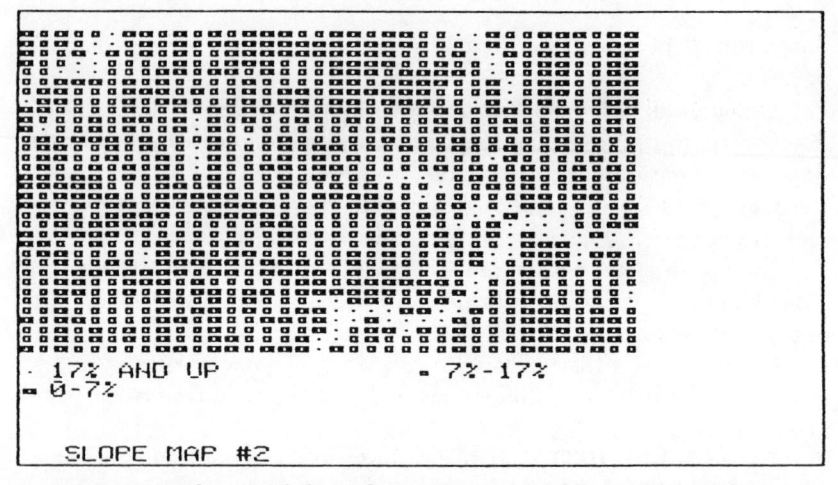

Figure 10-8. Hard copy of slope characteristics—Middle Gretna quadrangle.

The final factor considered was *noise*. Four categories of land were identified within this characteristic: none, little, moderate, and considerable.

Although all areas within this characteristic were likely to be usable for urban development, areas with the least noise were considered most suitable, and those with the most noise were considered least suitable.

Figures 10-16, 10-17, and 10-18 illustrate the noise characteristics in each study area.

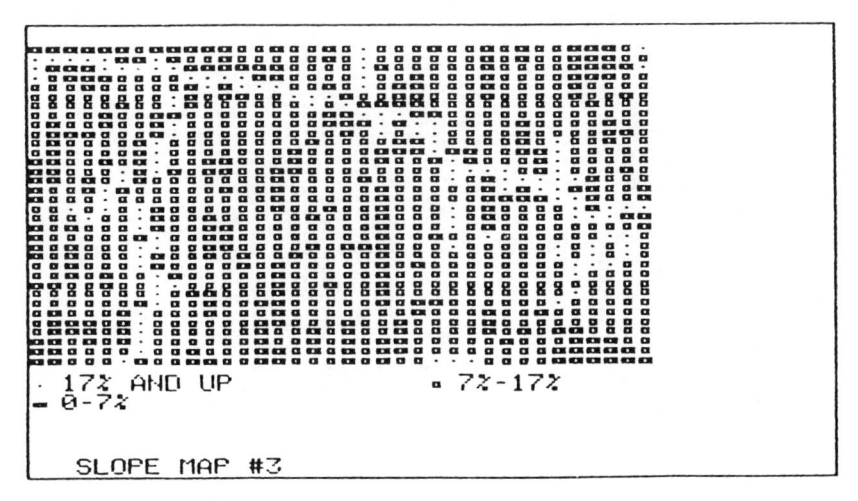

Figure 10-9. Hard copy of slope characteristics—Upper Gretna quadrangle.

In order to identify the areas that would qualify as "most suitable" in aggregate (based on the various considerations indicated above), researchers subjected each map to a process of prioritization of characteristics or groups of attributes, as well as prioritization of attributes within each group. A sliding scale of 0 to 1 was used to designate weights for different characteristics, while a sliding scale of 0 to 15 was used to designate weights for specific attributes within each group. Table 10-1 presents a summary of assigned weights.

The weight factors were assigned to each attribute by using the recode utility. Attributes that were assigned higher weight factors were recoded with colors bearing higher numerical code equivalents, and attributes that were assigned lower weight factors were recoded with colors bearing lower numerical code equivalents. The weight factors for each group of attributes (each map) took the form of overall multipliers.

The opaque overlay utility was invoked to combine all of the recoded and weighted maps, and the products of the overlay process became the "most suitable" alternative sites for urban use within each window.

A threshold of 12 was established as the minimum qualifying score for urban suitability. Land parcels (cells) displaying a score of 13 or higher were rated most suitable, while sites displaying a score of 12 were rated somewhat suitable. All other sites were deemed unsuitable. Figures 10-19, 10-20, and 10-21 illustrate the suitability of sites in each of the study areas.

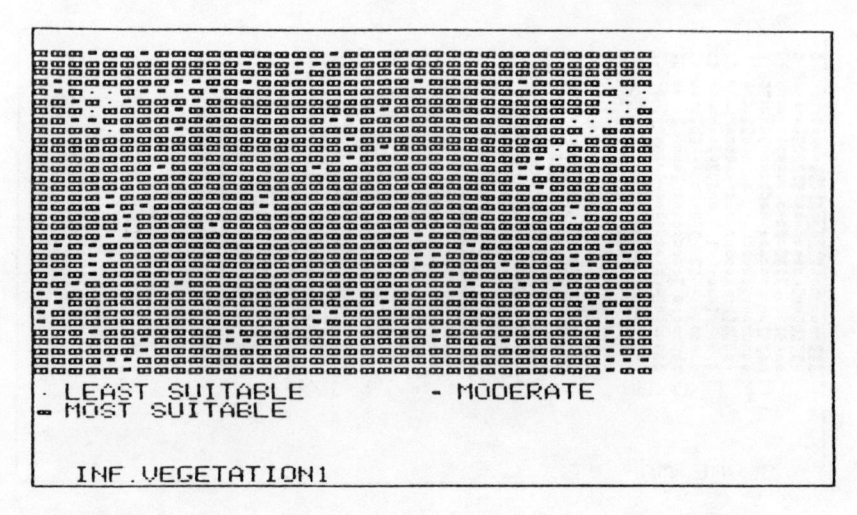

132 CARTOGRAPHY AND SITE ANALYSIS WITH MICROCOMPUTERS

Figure 10-10. Hard copy of vegetation cover—Lower Gretna quadrangle.

TABLE 10-1.

Characteristic	Wt.	Category	Wt.
Soil	1.00	Most suitable	14
		Marginally suitable	8
		Least suitable	2
Floodplain	0.65	Areas not subject to flooding	14
		Areas likely to be flooded	6
		Areas inundated	1
Slope	0.45	Slope < 7%	14
		7% < slope < 17%	8
		Slope > 17%	1
Vegetation	0.30	Little or no tree cover	13
		Moderate tree cover	6
		Heavy tree cover	1
Surface water	0.15	Dry land	14
		Intermittent streams	10
		Perennial streams	6
		Lakes, reservoirs, ponds, swamps, and marshes	1
Noise	0.04	None	14
		Little	9
		Moderate	4
		Considerable	0

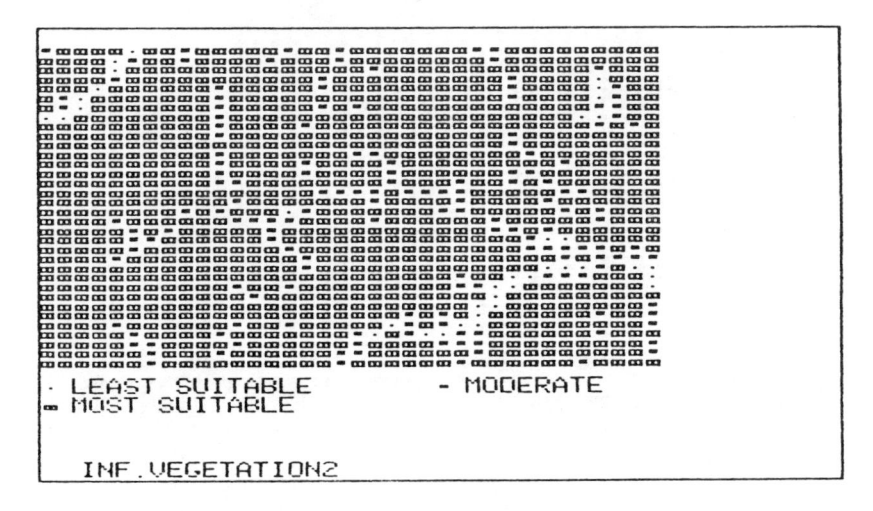

Figure 10-11. Hard copy of vegetation cover—Middle Gretna quadrangle.

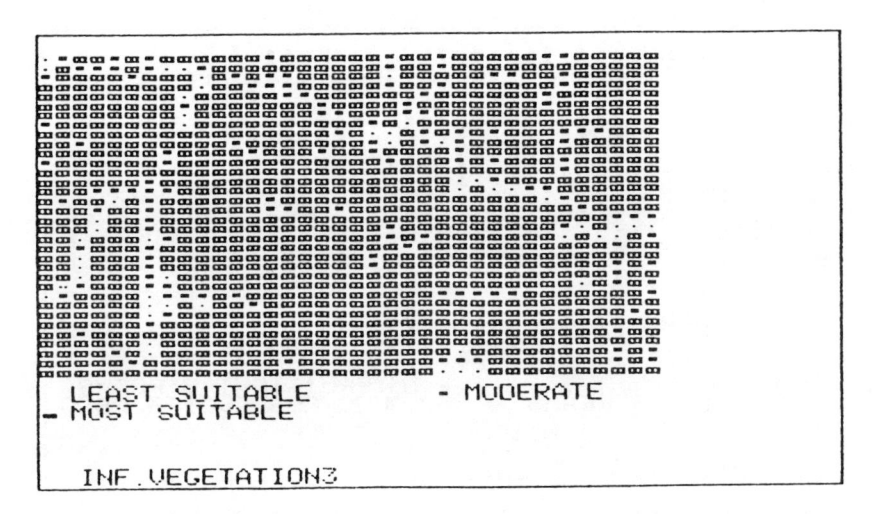

Figure 10-12. Hard copy of vegetation cover—Upper Gretna quadrangle.

The site configurations in each study area were evaluated by means of the evaluation utility. The resulting aggregate scores derived are shown in table 10-2.

It is very obvious from table 10-2 that the first alternative scored higher than either the second or the third alternative. This implies that

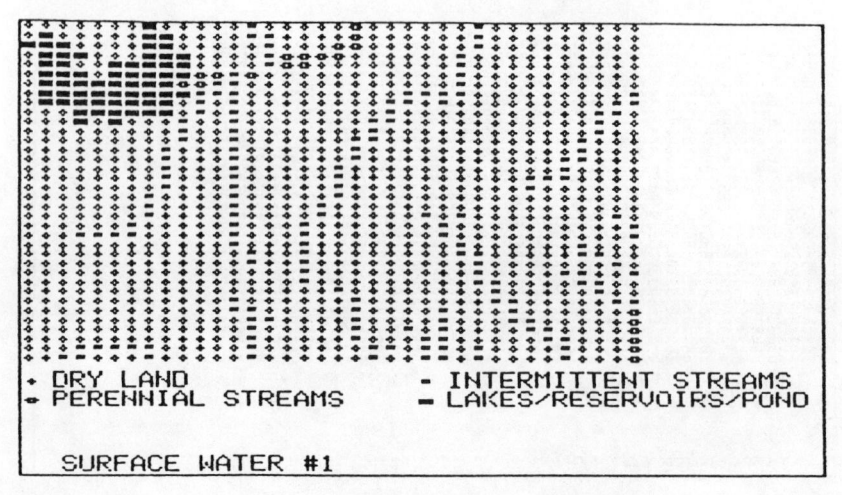

Figure 10-13. Hard copy of distribution of surface water—Lower Gretna quadrangle.

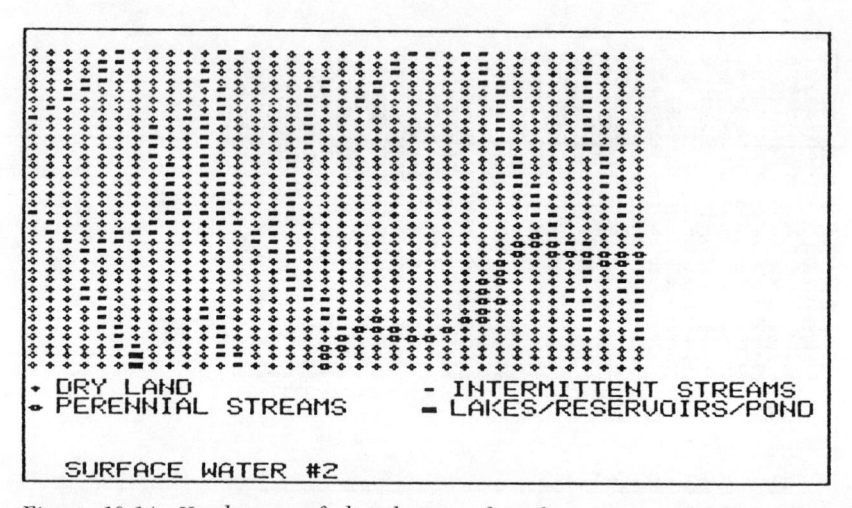

Figure 10-14. Hard copy of distribution of surface water—Middle Gretna quadrangle.

the first alternative had a larger share of cells that met the threshold value of 13.

The first alternative also showed the lowest standard deviation among the three alternatives. The interpretation of this value is that the cells within the first alternative varied less in their desirable attributes than

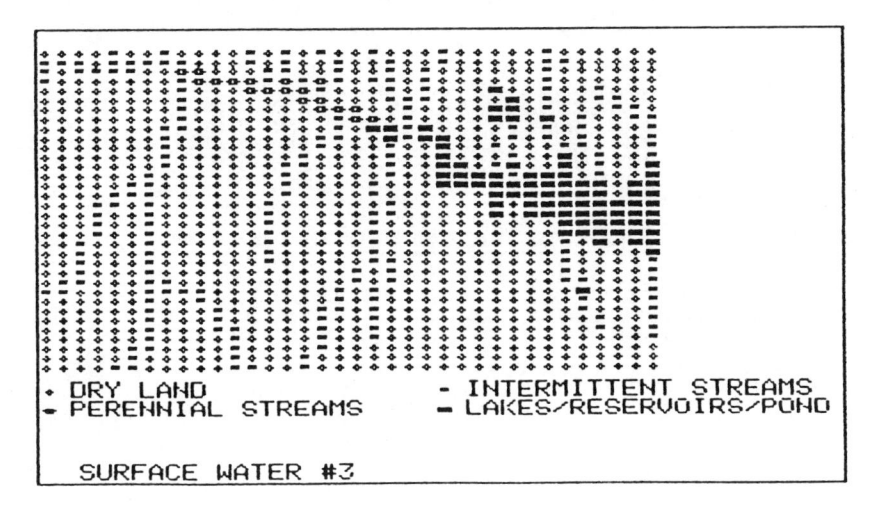

A Utility for Evaluating Alternative Sites 135

Figure 10-15. Hard copy of distribution of surface water—Upper Gretna quadrangle.

did those within the other two alternatives. Greater homogeneity was thus found in the attributes of cells within the first alternative than in those of cells within the other alternatives.

The possibility of securing a coherent land parcel represented by cells with scores equal to or above the threshold value in the second and third alternatives may be slight. Many contiguous cells with scores below the threshold value must be included in order to obtain a coherent land parcel. This difficulty is reduced in the first alternative—a point that is quite evident from the figures.

An integrated land parcel assembled from a contiguous group of cells with scores of 13 and above in the first alternative (the Lower Gretna area) would represent the "most suitable" area for urban use.

TABLE 10-2.

Alternative	Threshold Value	Standard Deviation	Score
Lower section	13	1.4	4,093
Middle section	13	2.7	3,746
Upper section	13	2.3	3,397

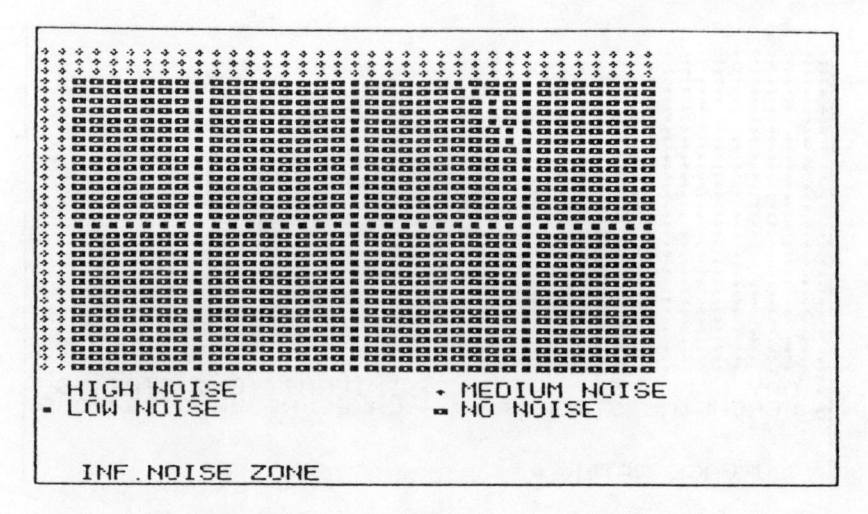

136 CARTOGRAPHY AND SITE ANALYSIS WITH MICROCOMPUTERS

Figure 10-16. Hard copy of noise characteristics—Lower Gretna quadrangle.

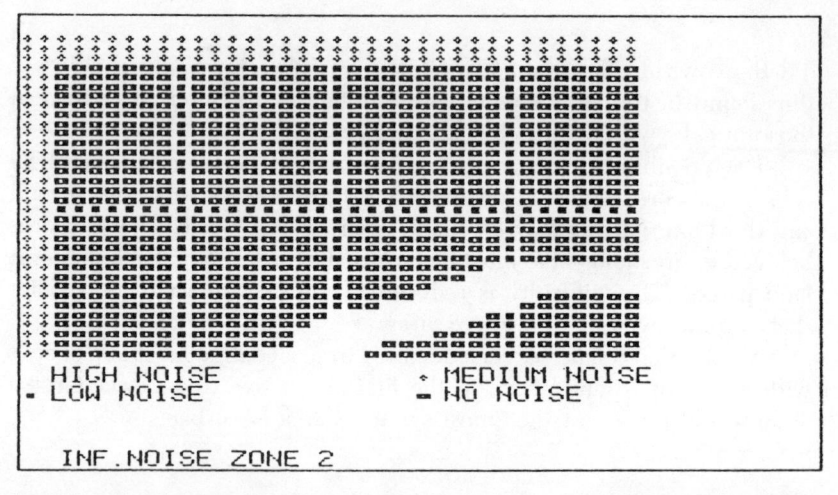

Figure 10-17. Hard copy of noise characteristics—Middle Gretna quadrangle.

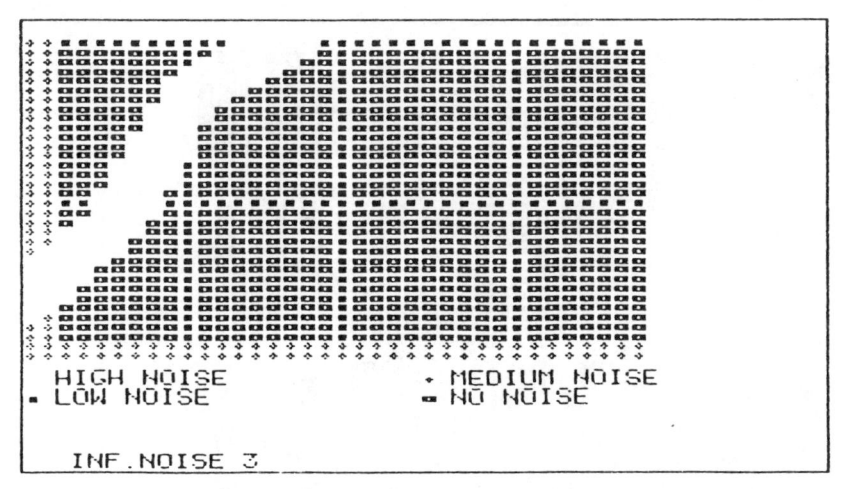

A Utility for Evaluating Alternative Sites 137

Figure 10-18. Hard copy of noise characteristics—Upper Gretna quadrangle.

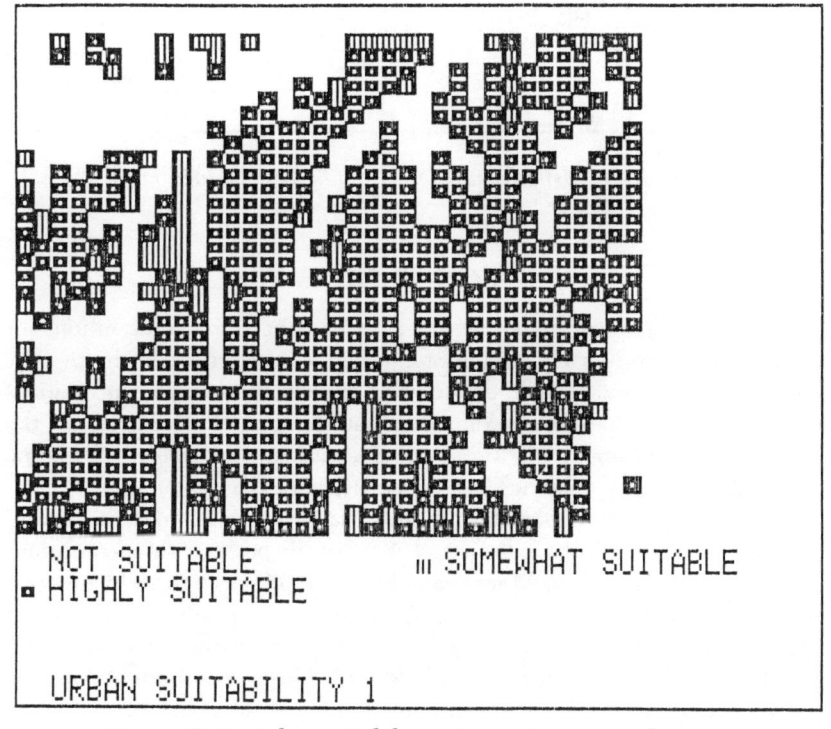

Figure 10-19. Urban suitability—Lower Gretna quadrangle.

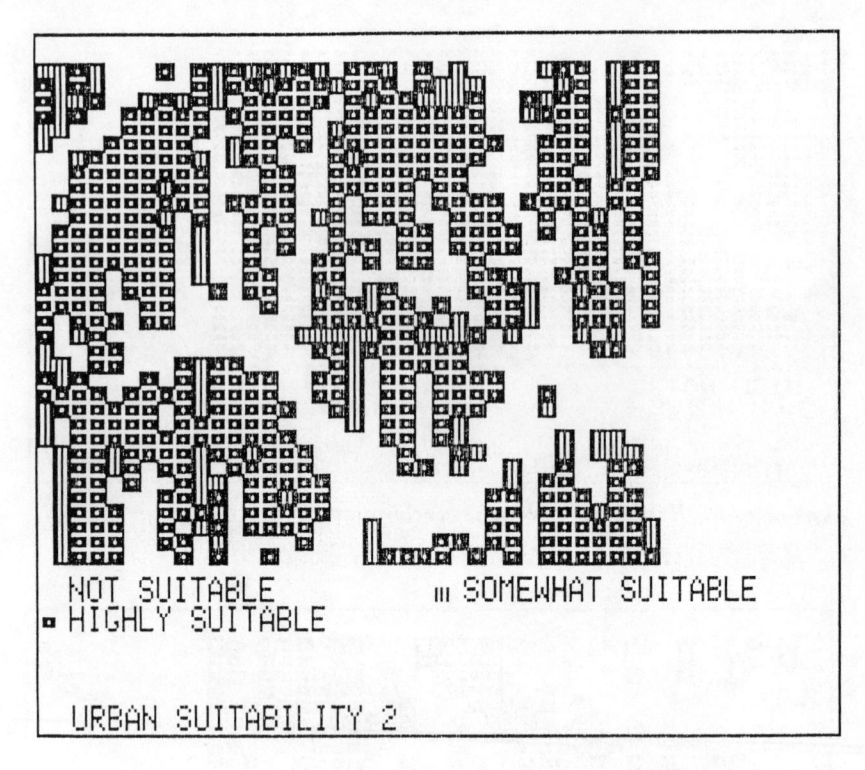

Figure 10-20. Urban suitability—Middle Gretna quadrangle.

CONCLUSION

The methodology suggested in this chapter is one of several options for site selection and for evaluation of alternatives. Variations of the overlay technique may be innovated and applied through appropriate combinations of utilities. Judiciousness in the comparative rating of attributes and the weighting of groups of attributes provides a justifiable means of finding a suitable site that matches the expectations of the user.

The demonstration study indicates that the process may not always yield a land parcel of contiguous cells that all possess ideal conditions for the desired purpose. More than likely, the process will yield a motley assembly of cells, scattered across the study area. Considerable judgment must then be exercised by the user in aligning and configuring a coherent land parcel. This configuration should include the maximum percentage of desirable cells relative to the total number of cells that must be included for the sake of contiguousness.

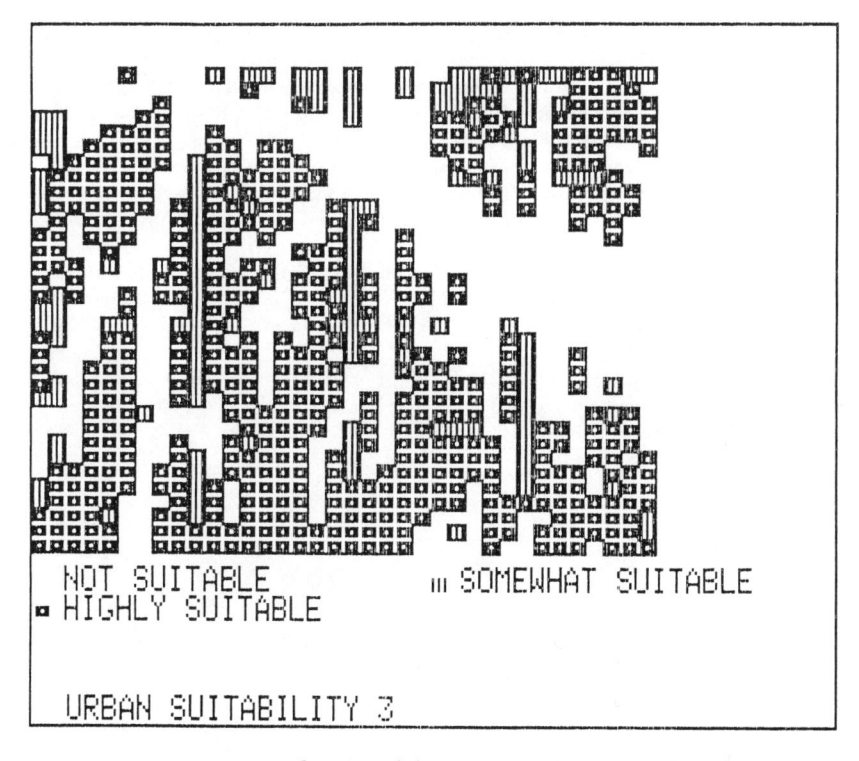

Figure 10-21. Urban suitability—Upper Gretna quadrangle.

The evaluation procedure provides a statistical means of examining and ranking several alternatives prior to making a final site selection. The score and its standard deviation provide the user with absolute and relative indicators for making satisfactory site selection and site location decisions. With a little creativity, the simple process designed and presented above can be refined and made more objective.

NOTES

1. *Alteration* in this context refers to any intentional change or modification performed on any attribute of a natural site. Alterations are not restricted to terrain modification, as the term *altered site* might imply; they may affect any attribute category—such as land use, density, or value—and may affect one or more attributes within a category.
2. Omaha–Council Bluffs Metropolitan Area Planning Agency, "Environmental Resource Analysis Program: A Resource Information Base" (Omaha: Omaha–Council Bluffs MAPA, 1979), p. 13.
3. A detailed discussion of urban use averages is contained in Omaha–Council Bluffs

Metropolitan Area Planning Agency, "Environmental Resource Analysis Program: A Resource Information Base" (Omaha: Omaha–Council Bluffs MAPA, 1979), pp. 35–41.

4. A detailed discussion of flooding is contained in Omaha–Council Bluffs Metropolitan Area Planning Agency, "Environmental Resource Analysis Program: A Resource Information Base" (Omaha: Omaha–Council Bluffs MAPA, 1979), pp. 66–68.

5. A detailed discussion of the selection of sites based on slope is contained in Omaha–Council Bluffs Metropolitan Area Planning Agency, "Environmental Resource Analysis Program: A Resource Information Base" (Omaha: Omaha–Council Bluffs MAPA, 1979), pp. 57–63.

6. A detailed discussion of the selection of vegetated areas for urban use is contained in Omaha–Council Bluffs Metropolitan Area Planning Agency, "Environmental Resource Analysis Program: A Resource Information Base" (Omaha: Omaha–Council Bluffs MAPA, 1979), pp. 50–57.

7. A discussion of surface water suitability ratings is contained in Omaha–Council Bluffs Metropolitan Area Planning Agency, "Environmental Resource Analysis Program: A Resource Information Base" (Omaha: Omaha–Council Bluffs MAPA, 1979), pp. 48–50.

ELEVEN

The Visibility Analysis Utility

The visibility analysis utility is an innovative tool that enables a user to determine omnidirectional visibility corridors around a selected viewing station of given elevation. The utility uses topographic maps generated by map creation and editing utilities.

The visibility analysis utility allows identification of desirable locations of vantage points for user-designated scenic views. Conversely, the utility allows discovery of undesirable vantage points, from which objectionable views (as designated by the user) may be visible. The purpose of the utility is to maximize the former and minimize the latter simultaneously.

THE PRINCIPLE OF VISIBILITY ANALYSIS

The principle by which the utility enables a user to determine whether a point B on a surface is visible from an observer's station O is graphically represented in figure 11-1.

The horizontal plane (represented as OA) is the reference plane of vision for the observer O. The first step is to select a line of vision— characterized in the drawing by the line from O (the observer station) to an elevated point B, which is being verified for visibility from point O. The tangent of B at O is then calculated with respect to the reference

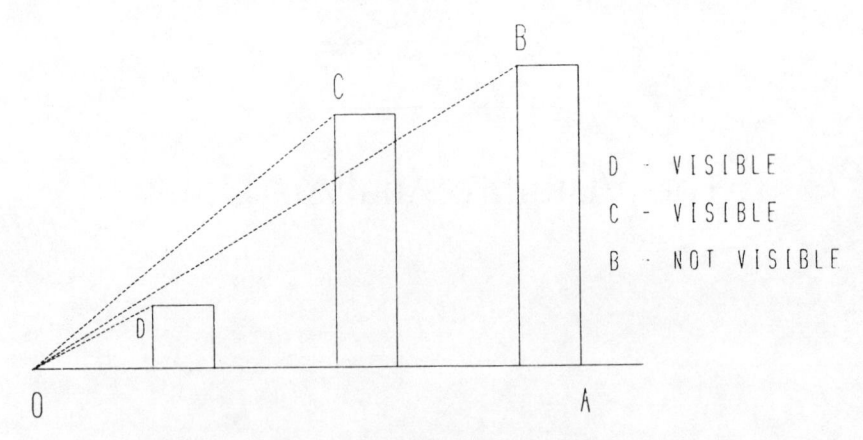

Figure 11-1. Theory of visibility.

plane OA. Similarly, the tangents of all intervening cells between the observer station O and point B (in this instance, C and D) are also calculated with respect to the reference plane OA. If the value of the tangent for any intervening cell exceeds that for point B, then B is not visible from observer station O.

The above procedure is repeated with reference to the observer station O for every cell contained in the viewing window. Only visible cells are selected for display.

In the diagram in figure 11-1, point B is not visible from O because the tangent of C at O exceeds the tangent of B at O. Point C is visible from O, however, in spite of D's intervening between O and C. Visibility is possible because the tangent of D at O is less that that of C at O.

THE PROCEDURE FOR USING THE VISIBILITY ANALYSIS UTILITY

The user first selects the topographic map on which the visibility analysis is to be conducted, loading it onto the screen by means of the load utility.

The next task is to identify and locate the viewing station from which omnidirectional views are to be analyzed on this map. The elevation height of the viewing station must be taken into account and designated prior to invoking the visibility analysis utility. If the elevation of the viewing station differs from the elevation of the prevailing topography at that location, the editing utility should be invoked to modify the color of the corresponding cell, in order to reflect the difference in elevation.

The Visibility Analysis Utility

This map then serves as the input source map for conducting the visibility analysis procedure.

The third task is to activate the visibility analysis utility and designate the "viewing window." The viewing window represents an area envelope of part of the displayed map, within which the visibility analysis from the selected viewing station is considered feasible. The total spatial extent of the display may serve as the viewing window. This definition of *viewing window* is particularly relevant in cases where the topography is relatively flat and even and visibility of very distant features is obscured as a result of the earth's curvature.

The product map derived from the application of this utility displays all cells that are visible from the viewing station and suppresses all cells that are not visible.

Several iterations of the above process must be attempted before the best selection of vantage points is obtained. Each iteration may include some simple modifications and enhancements (as suggested below) toward maximizing the best views and minimizing the most undesirable ones. First, the view from a viewing station may be enhanced by varying the viewing station's elevation. The visibility analysis may then be conducted using the viewing station at its modified elevation. Second, low spots may be elevated, or the elevations of obstructions may be lowered by means of suitable terrain modifications. Such modifications may either hide objectionable views or open up new scenic views.

In either instance, the user should transfer the proposed terrain variations onto the topography map prior to conducting the visibility analysis. The modifications themselves become themes for estimating earthwork excavation or filling, using the cut-and-fill utility described in chapter 12.

The following subroutine enables a user to activate the visibility analysis utility:

```
13000 REM *** SUBROUTINE VISIBILITY ANALYSIS ***
13010 REM THIS SUBROUTINE PERFORMS VISIBILITY ANALYSIS
13020 PRINT D$;"PR#0"
13030 HOME: PRINT "<< TOP LEFT CORNER OF WINDOW >>"
13035 PRINT D$;"PR#";SL: PRINT "N"
13040 GOSUB 1000: IF F2% = 1 THEN 13381
13050 GOSUB 1200: IF F1% = 1 THEN 13040
```

```
13060 TX = A: TY = B: REM SAVE TOP LEFT
      COORDINATES
13065 PRINT D$;"PR#0"
13070 HOME: PRINT "<< BOTTOM RIGHT CORNER OF
      WINDOW >>"
13075 PRINT D$;"PR#";SL: PRINT "N"
13080 GOSUB 1000: IF F2% = 1 THEN 13381
13090 GOSUB 1200: IF F1% = 1 THEN 13080
13100 BX = A: TY = B: REM SAVE BOTTOM
      RIGHT COORDINATES
13105 PRINT D$;"PR#0"
13110 HOME: PRINT "<< OBSERVER STATION >>"
13115 PRINT D$;"PR#";SL: PRINT "N"
13120 GOSUB 1000: IF F2% = 1 THEN 13381
13130 GOSUB 1200: IF F1% = 1 THEN 13120
13140 CX = A: CY = B: REM SAVE OBSERVER
      STATION COORDINATES
13150 IF TX > BX OR TY > BY OR CX < TX OR CX >
      BX OR CY < TY OR CY > BY THEN 13030: REM
      ILLEGAL ENTRY
13180 IF TX = 0 OR TY = 0 OR BX = R4% OR BY =
      R4% THEN 13200
13184 COLOR = 15: REM PLOT WINDOW BOUNDARY
13186 HLIN TX - 1,BX + 1 AT TY - 1
13188 HLIN TX - 1,BX + 1 AT BY + 1
13190 VLIN TY - 1,BY + 1 AT TX - 1
13195 VLIN TY - 1,BY + 1 AT BX + 1
13198 COLOR = 0: PLOT TX - 1,CY: PLOT CX,BY +
      1
13210 FOR  J = TY TO BY
13220     FOR I = TX TO BX
13230         COLOR = 0: PLOT I,J
13250         A = I - CX: B = J - CY: REM
              DRAW IMAGINARY LINE TO
              CHECK FOR VISIBILITY
13260         L = SQR (A^2 + B^2)
13270         IF L = 0 THEN 13360
13280         UX = A/ L: UY = B / L: REM
              CALCULATE UNIT VECTOR
13290         T = (D%(I,J) - D%(CX,CY)) /
              INT (L): REM CALCULATE SLOPE
13300         K = R1%
```

```
13310              IF K > L THEN 13360: REM WHILE
                     LOOP
13320                A1 = INT (.5 + CX + K *
                     UX)
13330                B1 = INT (.5 + CY + K *
                     UY)
13340                IF (D%(A1,B1) -
                     D%(CX,CY)) / K > T THEN
                     13370: REM NOT VISIBLE
13350              K = K + 1 :   GOTO 13310:
                     REM END OF WHILE LOOP
13360              COLOR = D%(I,J): PLOT I,J
13370      NEXT
13380 NEXT
13381 PRINT D$;"PR#0"
13382 PRINT "WOULD YOU LIKE TO SAVE THE
      MAP ?";
13383 GOSUB 1400: PRINT
13384 IF A$ = "N" THEN 13390
13385 GOSUB 23000: REM SAVE THE MAP
13386 GOSUB 970: REM REDRAW THE MAP
13390 PRINT D$;"PR#";SL: PRINT "N"
13390 RETURN
```

Lines 13020 to 13130 seek the boundary coordinates for the viewing window and those for the observer station. The loop of lines 13210–13380 implements the visibility analysis algorithm, the principle of which has already been described.

TYPICAL EXAMPLES OF THE USE OF THE VISIBILITY ANALYSIS UTILITY

The following demonstration exercise was designed to identify prime building sites possessing the best visibility of a reservoir.

Two methods could be used: the user could locate viewing stations at each cell around the reservoir and verify the reservoir's visibility in each case; or the user could work on the reverse logic that any site visible from the reservoir has a view of the reservoir. The latter alternative is by far the more efficient.

The starting point of this analysis was to generate a contour map using the map creation and editing utilities. This map is shown in figure

11-2. A three-dimensional terrain model of the site is shown in figure 11-3.

The next step in the process was to locate the viewing window and to identify the location and elevation of the viewing station. In this case, the elevation of the viewing station did not have to be altered manually, since the intention was to observe whether or not the surface of the reservoir waters (and not any stations above or below it) was visible.

Figure 11-4 displays the locations of all areas visible from the selected viewing station. The blank areas represent areas not visible from the selected viewing station. It may be concluded that building sites corresponding to the displayed cells within the viewing window are prime properties in the study area.

Of course, as buildings are built on the cells in the foreground, the views from lots in the background tend to be obstructed. In order to maximize the number of lots with views, the designer would have to stagger buildings in such a manner that visibility from lots in the rear was not adversely affected by buildings closer to the water. Another (and far more expensive) alternative might be to alter the terrain in the rear through a terracing operation. Buildings on lots in the rear might then

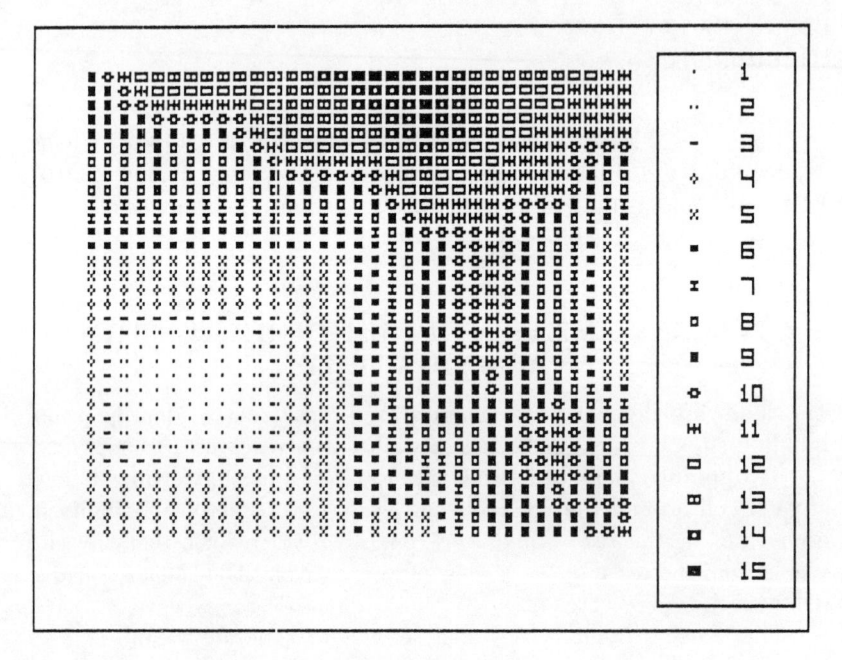

Figure 11-2. Contour map of the reservoir site.

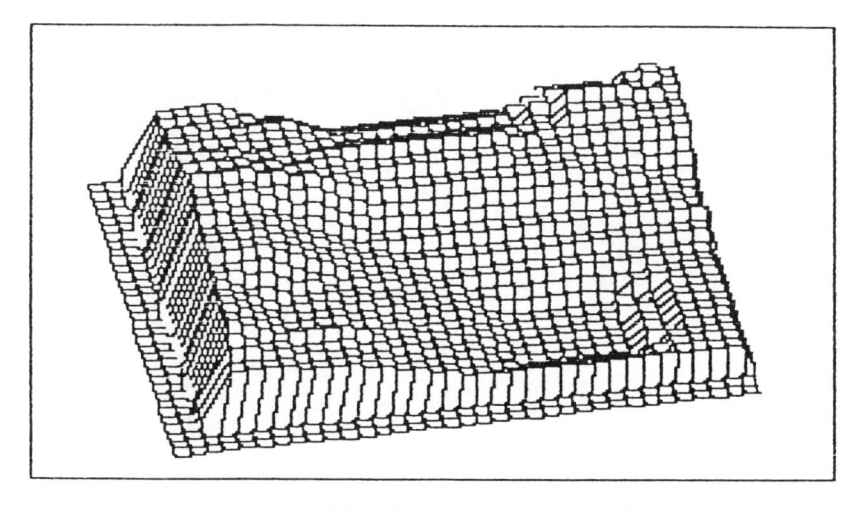

Figure 11-3. Terrain model of the reservoir, as viewed from the west.

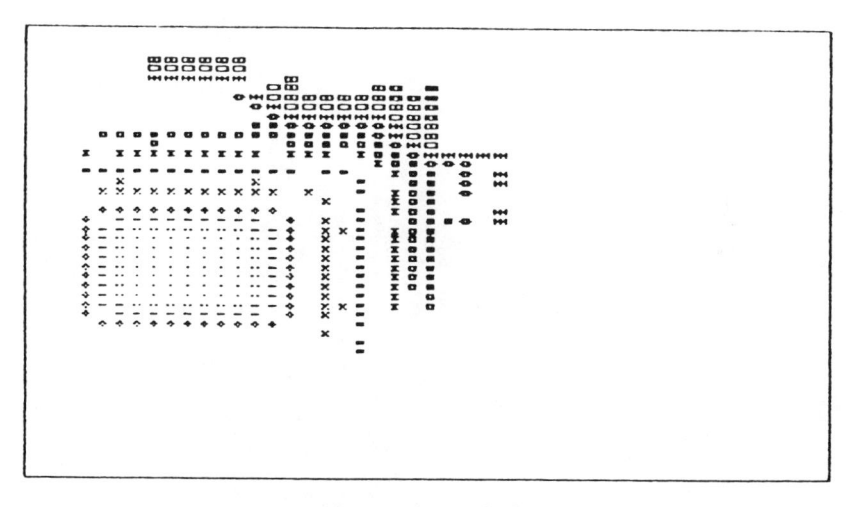

Figure 11-4. Location of areas from which the reservoir is visible.

secure an elevated viewing corridor of the reservoir, over the tops of buildings on intervening lots.

CONCLUSION

Topography plays a crucial role in endowing certain locations with spectacular views. Topographic and other features of the built environment tend to enhance scenic views in some instances and to impede them in others. The visibility analysis utility enables a user to determine locations that are suited for adoption as vantage points with desirable views. It also enables a user to analyze the impact of various topographic alterations on views from different viewing stations, facilitating judicious site modifications.

TWELVE

Terrain Modification and the Cut-and-Fill Utility

Estimating the earthwork involved in terrain modification is a common task undertaken by planners, designers, and landscape architects at the time of layout preparation. The volumes of earth to be removed from some locations and those to be deposited at other locations must be known in advance, to allow estimation of the impending need for importing or exporting earth to or from the site.

The principal objective of terrain modification is to achieve a desirable terrain for a specific purpose. Terrain modification always involves some cutting and some filling. The ideal balance in cutting and filling is achieved when the desired formation is obtained through removal of a volume of earth that is exactly equal to the volume of earth that needed to be deposited; in such an instance, no import or export of earth outside the building site is required in order to achieve the desired formation. Still, this may not always be practically feasible—although it may be used as a goal to strive toward in any case.

THE OBJECTIVE OF THE CUT-AND-FILL UTILITY

The principal objective of the cut-and-fill utility is to estimate the volumes of cutting and filling that may be required in order to achieve a desired formation of terrain. The balancing of cut-and-fill, as described above, is treated as a special optimization option in the utility.

SOME BASIC RULES FOR VALIDITY, ACCURACY, AND RELIABILITY OF CUT-AND-FILL CALCULATIONS

In order to estimate cut-and-fill for any area, the user must have the topography map of natural terrain that is generated by using the map creation and editing utilities. Considerable care must be exercised in selecting the scale for the preparation of this base map. The scale should maximize the validity, accuracy, and reliability of the cut-and-fill utility's results.

Cut-and-fill estimations should be carried out only on very detailed topography maps because the level of error increases as the scale increases. For instance, the errors in the results of cut-and-fill estimation reach unreasonable proportions (bordering on the absurd) if the utility is used on a source map bearing a scale of 4 inches = 1 mile.

A valid, accurate, and reliable estimation of cut-and-fill earthwork can be ensured if a reasonable correlation exists between the areal dimensions of each of the low-resolution graphics cells (that is, the length and breadth of the cell) and the vertical dimensions of each, as abstracted by the color code (the elevation of the cell). The calibration process embedded in the cut-and-fill utility uses "compatible" areal and vertical units (for example, square feet and feet; acres and feet would be considered incompatible).

A topography map prepared in accordance with the above principle becomes the base map for performing cut-and-fill operations.

The user may graphically simulate desired modifications and alterations to the natural terrain displayed in the above map by using the editing utilities, and then saving the modified drawing by means of the save utility. Each modified topography map serves as one of a possible series of proposed land formation alternatives. A number of such alternatives may be generated by the user, through the above method.

THE PROCEDURE OF CUT-AND-FILL ESTIMATION

Two maps serve as input data for cut-and-fill estimation: These are the topography map for natural terrain, and the topography map for altered terrain. First, the natural terrain map must be loaded by means of the load utility. The cut-and-fill utility may then be activated, the name of the altered terrain map being supplied by the user in response to the first query.

The first step is the designation of a window boundary within which the earthwork estimation is to be conducted. The next step is calibration; the numerical value of the contour interval and the linear scale used in the map are the parameters used for this purpose. Both parameters should be expressed in compatible units, as described earlier. Cut-and-fill volumes are estimated, and the results are then displayed graphically, based on the following algorithm.

ALGORITHM FOR CUT-AND-FILL ESTIMATION AND GRAPHIC REPRESENTATION

Cut-and-fill volume is estimated by comparing the map displaying the altered terrain with the base map of unaltered (natural) terrain. The numerical values corresponding to the colors of each cell on the altered terrain map are subtracted from the numerical values corresponding to the colors of the identical cell on the unaltered terrain map. The results of subtraction for each cell are multiplied by the area of the cell to identify the volume of earthwork required for that particular cell.

Three results are possible as a consequence of the subtraction: a result of zero, which is interpreted as NO-CUT-NO-FILL at that cell; a positive result, which is interpreted as FILL at that cell; or a negative result, which is interpreted as CUT at that cell.

These results serve as the numerical codes for generating a series of displays. One display shows the locations of all cells that are to be subjected to NO-CUT-NO-FILL cutting and filling. This display is generated using the color red to represent cut, the color light blue to represent fill, and the color gray to represent NO-CUT-NO-FILL.

All positive and negative results from the subtraction process are aggregated, independent of one another, to derive statistics on the volumes of filling and of cutting required. A statistical profile of the cut-and-fill volumes required to alter the natural terrain is also derived for display.

A second display shows the intensities of fill at locations where filling has to be performed. Intensities are represented by color codes; the numerical value of the color code at any cell location can be derived by figuring the difference between the numerical color code of that cell in the contour map of natural terrain and the corresponding numerical color code in the contour map of altered terrain. The higher the numerical value of a cell's color code, the greater the intensity of fill.

A third display shows the intensities of cutting at locations where cutting has to be performed. The same principle used for estimating fill intensities is used for this representation.

PROCEDURE OF CUT-AND-FILL OPTIMIZATION

As noted previously, an optimal solution for cutting and filling is one in which the volume of cutting is balanced or almost balanced by the volume of filling, in the course of creating the desired terrain formation. This solution is subject to certain constraints. The highest and lowest permissible elevations anywhere on the site must be specified by the user.

The optimization option of the cut-and-fill utility makes iterative adjustments and readjustments to the natural formation in order to equalize the volumes of cutting and filling—subject to a maximum permissible equalization error of 10 percent. This procedure results in the generation of a series of displays and statistics corresponding to the "optimum terrain formation."

THE ALGORITHM FOR CUT-AND-FILL OPTIMIZATION

The algorithm that follows is used for securing the optimum terrain formation.

Step 1. A maximum ten trials of step 2 are conducted. Either volumes of cut and fill are equaled within a tolerance limit of 10 percent in the course of those trials, or the utility gives up.

Step 2. Steps 2.1 and 2.2 are repeated for every cell within the boundary window.

Step 2.1. If the volume of fill is greater than the volume of cut, and if the numerical color code of the cell in the modified formation is greater than that of the corresponding cell in the natural terrain, and if the numerical color code of that cell is greater than the numerical color code for the lowest permissible elevation, then the elevation of the cell in the formation map should be decreased.

Step 2.2. If the volume of cut is greater than volume of fill, and if the numerical color code of the cell in the modified formation is less than that of the corresponding cell in the natural terrain, and if the numerical color code of that cell is less than the numerical color code for the highest permissible elevation, then the elevation of the cell in the formation map should be increased.

This option uses a trial-and-error method to reduce the amount of manual effort required to achieve a balance between the volume of cut and the volume of fill. This does not necessarily produce a solution that represents the ideal terrain alteration for the site. Some manual adjustments may still have to be made in order to derive a solution more acceptable under the circumstances than the "optimal" solution.

The following subroutine enables a user to perform the cut-and-fill utility:

Terrain Modification: Cut-and-Fill Utility

```
14000 REM *** SUBROUTINE CUT AND FILL ***
14005 REM   THIS SUBROUTINE PERFORMS CUT-AND-
           FILL COMPUTATION
14010 REM   AND FINDS LEVELS AT WHICH CUT AND
           FILL ARE BALANCED
14020 PRINT D$;"PR#0": PRINT D$;"IN#0"
14030 HOME: PRINT "NAME OF FORMATION CONTOUR
      MAP ";
14040 INPUT NN$
14050 IF NN$ = " " THEN 14690
14060 GOSUB 1300: PRINT
14070 NN$ = "MAP." + NN$
14080 PRINT D$;"BLOAD";NN$;",A1024,D";DR$
14085 PRINT D$;"PR#0"
14090 HOME :   PRINT "TOP LEFT BOUNDARY":   REM
      SEEKS WINDOW COORDINATES
14100 PRINT D$;"PR#";SL: PRINT "N"
14105 PRINT D$;"IN#";SL
14110 GOSUB 1000: IF F2% = 1 THEN 14690
14120 GOSUB 1200: IF F1% = 1 THEN 14110
14130 TX = A: TY = B
14140 PRINT D$;"PR#0"
14150 HOME: PRINT "BOTTOM RIGHT BOUNDARY"
14160 PRINT D$;"PR#";SL: PRINT "N"
14170 GOSUB 1000: IF F2% = 1 THEN 14690
14180 GOSUB 1200: IF F1% = 1 THEN 14170
14190 BX = A: BY = B
14200 IF TX > BX OR TY > BY THEN 14085: REM
      ILLEGAL ENTRY
14210 GOSUB 1500:   REM SAVE THE DISPLAY WINDOW
      IN MATRIX E%
14220 REM PERFORM SCALING OF CELL
14230 PRINT D$;"PR#0"
14240 HOME: PRINT "PRESS STYLUS AT THE
      BEGINNING OF THE SCALE"
14250 PRINT D$;"PR#";SL: PRINT "N"
14260 GOSUB 1000: IF F2% = 1 THEN 14690
14270 MX = X: MY = Y
14280 PRINT D$;"PR#0"
14290 HOME: PRINT "PRESS STYLUS AT THE END OF
      THE SCALE"
14300 PRINT D$;"PR#";SL: PRINT "N"
```

```
14305 GOSUB 1000: IF F2% = 1 THEN 14690
14310 NX = X: NY = Y
14320 PRINT D$;"PR#0": PRINT D$;"IN#0"
14330 INPUT "NUMBER OF UNITS ";A$
14340 LM = VAL (A$)
14350 INPUT "NAME OF UNITS ";NU$
14360 K$ = NU$: NU$ = " CUBIC " + LEFT$(NU$,7)
14370 HOME: INPUT "INITIAL ELEVATION VALUE ";
      A$
14380 ZZ = VAL(A$) - 1
14390 INPUT "NUMBER OF UNITS BETWEEN EACH
      COLOR CODE ";A$
14400 QU = VAL (A$) : IF QU = 0 THEN 14390
14410 REM PERFORM SCALING OF CELLS
14420 L = SQR ((MX - NX)^2 +  (MY - NY)^2):
      REM LENGTH OF INPUT SCALE
14430 IF L = 0 THEN 14240: REM ILLEGAL ENTRY
14440 U = LM / L: REM CALCULATE LENGTH OF EACH
      UNIT
14450 UX = U : REM X OF CELL
14460 UY = U * (4 / 7): REM Y OF CELL
14470 V = UX * UY * QU: REM VOLUME OF CELL
14480 REM END OF SCALING CELL
14490 GOSUB  14700:   REM  PERFORM  CUT-AND-
      FILL COMPUTATION
14495 HOME:   PRINT  "WOULD  YOU  LIKE
      OPTIMUM FORMATION MAP (Y/N) ?";
14500 GOSUB 1400: PRINT
14505 IF  A$  =  "N"  THEN 14690
14510 PRINT "HIGHEST POSSIBLE FORMATION
      ELEVATION"
14515 PRINT  "(";  ZZ  ;  "-"  ;  INT (ZZ +
      15  * QU);K$;") ";
14520 INPUT A$: HD = VAL (A$) - ZZ
14525 PRINT  "LOWEST POSSIBLE FORMATION
      ELEVATION NOT LESS THAN ";ZZ;" ";
14530 INPUT A$: LD = VAL(A$) - ZZ
14535 HOME : PRINT "CALCULATING . . ."
14540 HD = HD / QU: LD = LD / QU
14542 TRY = 0
14545 REM LOOP TILL CUT AND FILL ARE BALANCED
14546 FI = 0: CU = 0
```

```
14548 FOR J = TY TO BY
14550     FOR I = TX TO BX
14555         IF E%(I,J) > D%(I,J) THEN FI =
              FI + E%(I,J) - D%(I,J): GOTO
              14564
14560         IF E%(I,J) < D%(I,J) THEN CU =
              CU + D%(I,J) - E%(I,J)
14564     NEXT
14568 NEXT
14570 IF CU <= FI + AV AND CU >= FI - AV OR
      TRY > 10 THEN 14630:  REM STOP IF
      BALANCED WITHIN 10% OR 10 TRIALS ARE
      COMPLETED
14575 IF CU < FI THEN 14605
14580 FOR   J = TY TO BY: REM ELEVATE
      FORMATION
14585     FOR I = TX TO BX
14588         IF E%(I,J) < D%(I,J) AND E%(I,J)
              < HD THEN E%(I,J) = E%(I,J) + 1
14590     NEXT
14595 NEXT
14600 GOTO 14625
14605 FOR J = TY TO BY : REM DEPRESS FORMATION
14610     FOR I = TX TO BX
14612         IF E%(I,J) > D%(I,J) AND E%(I,J)
              > LD THEN E%(I,J) = E%(I,J) - 1
14615     NEXT
14620 NEXT
14625 TRY = TRY + 1: GOTO 14545: REM END LOOP
14630 GOSUB 14980: REM CLEAR SCREEN
14635 PRINT D$;"PR#0": PRINT D$;"IN#0": HOME
14640 PRINT "OPTIMUM FORMATION MAP"
14650 FOR J = TY TO BY
14655     FOR I = TX TO BX
14660         COLOR = E%(I,J): PLOT I,J
14665     NEXT
14670 NEXT
14675 PRINT "DO YOU WISH TO SAVE THE MAP
      (Y/N) ?";
14680 GOSUB 1400: PRINT
14685 IF A$ = "Y" THEN GOSUB 23000
14688 GOSUB 14700
```

156 CARTOGRAPHY AND SITE ANALYSIS WITH MICROCOMPUTERS

```
14690 PRINT: HOME
14891 XM = DC%: GOSUB 21000:  GOSUB 970  :
      REM REDRAW BASE MAP
14692 PRINT D$;"PR#";SL: PRINT "N": PRINT
      "IN#";SL
14695 RETURN
14699 REM
14700 REM ***   SUBROUTINE 14700 ***
14701 REM   THIS   SUBROUTINE PERFORMS CUT-AND-
      FILL COMPUTATIONS, AND    DISPLAYS  CUT-
      AND-FILL INTENSITIES
14705 CU = 0: FI = 0
14710 PRINT   D$;"PR#";SL: PRINT "N":    PRINT
      D$;"IN#";SL
14720 GOSUB 14980: REM CLEAR SCREEN
14730 FOR J = TY TO BY: REM CALCULATE VOLUME
      OF CUT AND FILL
14740     FOR I = TX TO BY
14750         IF E%(I,J) > D%(I,J) THEN FI =
              FI + E%(I,J)   - D%(I,J):
              COLOR  =   1: PLOT I,J:
              GOTO 14780:  REM CHECK EVERY
              CELL AND CALCULATE AMOUNT  OF
              FILLING AND DISPLAY IT IN RED
14760         IF E% (I,J) = D%(I,J) THEN
              COLOR = 5: PLOT I,J:  GOTO
              14780:   REM IF THERE IS  NO
              NEED FOR CUTTING OR  FILLING,
              DISPLAY IT IN GRAY
14770         CU   =   CU + D%(I,J)    -
              E%(I,J): COLOR = 7:  PLOT I,J:
              REM CALCULATE AMOUNT OF CUTTING
              AND DISPLAY IT IN LIGHT BLUE
14780     NEXT
14790 NEXT
14800 AV = .1 * ( CU + FI )/ 2 :   REM USED AS
      BENCHMARK FOR OPTIMIZATION
14805 PRINT D$;"PR#0": PRINT D$;"IN#0"
14810 PRINT "VOLUME OF CUT ";INT (100 * CU *
      V)/100;NU$
14815 PRINT "VOLUME OF FILL";INT (100 * FI *
      V)/100;NU$
```

```
14820 PRINT " DO YOU WISH TO SAVE THE MAP
      (Y/N) ?";
14825 GOSUB 1400: PRINT
14830 IF A$ = "Y" THEN GOSUB 23000
14834 PRINT D$;"PR#0": HOME: PRINT "FILL
      INTENSITY"
14835 PRINT D$;"PR#";SL: PRINT "N": PRINT
      D$;"IN#";SL
14840 GOSUB 14980: REM CLEAR DISPLAY
14845 FOR J = TY TO BY
14850     FOR I = TX TO BX
14855         IF E%(I,J) > D%(I,J) THEN
              COLOR = ABS (E%(I,J) -
              D%(I,J)): PLOT I,J
14870     NEXT
14875 NEXT
14880 PRINT D$"PR#0": PRINT D$;"IN#0"
14885 PRINT"DO YOU WISH TO SAVE THE MAP
      (Y/N) ?";
14890 GOSUB 1400: PRINT
14895 IF A$ = "Y" THEN GOSUB 23000
14900 PRINT D$;"PR#0": HOME: PRINT "CUT
      INTENSITY"
14905 PRINT D$;"PR#";SL: PRINT "N": PRINT
      D$;"IN#";SL
14910 GOSUB 14980: REM CLEAR DISPLAY
14915 FOR J = TY TO BY
14920     FOR I = TX TO BX
14925         IF E%(I,J) < D%(I,J) THEN
              COLOR = ABS (E%(I,J) -
              D%(I,J)): PLOT I,J
14935     NEXT
14940 NEXT
14945 PRINT D$"PR#0": PRINT D$;"IN#0"
14950 PRINT"DO YOU WISH TO SAVE THE MAP (Y/N)
      ?";
14955 GOSUB 1400: PRINT
14958 IF A$ = "Y" THEN GOSUB 23000
14960 PRINT D$;"PR#0" : PRINT D$;"IN#0"
14965 RETURN
14975 REM
14980 REM *** SUBROUTINE 14980 ***
```

```
14981 REM
14982 COLOR = 0
14984 FOR I = R TO R4%
14986     FOR J = R TO R4%
14988         PLOT I,J
14990     NEXT
14992 NEXT
14994 RETURN
```

Lines 14085–14200 allow selection of the window boundary. Lines 14220–14480 perform the calibration of the map. Subroutine 14700 is used by the cut-and-fill utility to compute the volume of cut and fill and to display cut-and-fill intensities.

The loop of lines 14545–14625 performs the cut-and-fill optimization procedure, consistent with the above algorithm.

Subroutine 14980 is used by the cut-and-fill utility to clear the display.

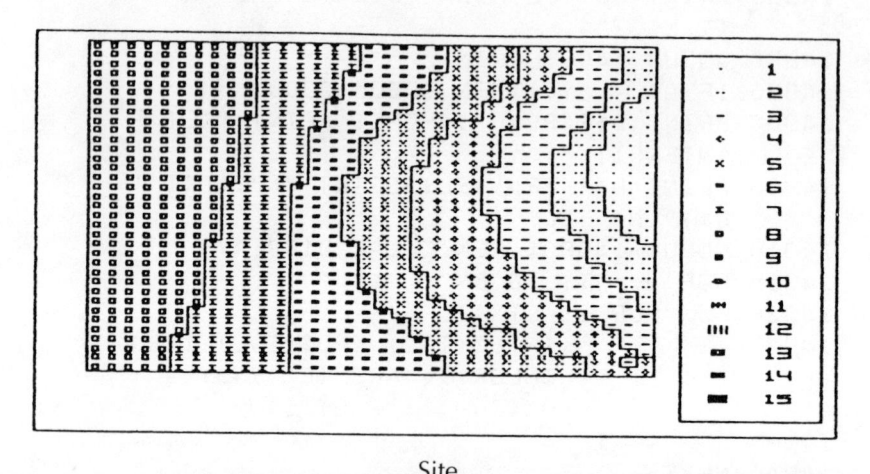

Site

LEGEND	
#	Key
1	82 feet
2	83 feet
3	84 feet
4	85 feet
5	86 feet
6	87 feet
7	88 feet

Figure 12-1. Hard copy of contours of the lot in South Lincoln.

A TYPICAL APPLICATION OF THE CUT-AND-FILL UTILITY

A demonstration study was undertaken to review terrain alterations needed for locating and constructing an apartment complex on a residential lot in South Lincoln, Nebraska. Contour information was available for the site at 1-foot intervals.

The preliminary step consisted of preparing the base map that showed the natural terrain of the lot, using the map creation and editing utilities (see figure 12-1). The topography map was used as the base map for creating alternative terrain modifications, and then was saved for future retrieval by means of the save utility.

A sketch layout of buildings was prepared independently to determine the preferred locations of proposed buildings on the lot. Modifications of the natural terrain that would render it capable of siting buildings were suggested. The proposed modifications were incorporated into the topographic base map, and this was then saved by means of the save utility. The map displaying the terrain modifications is shown in figure 12-2.

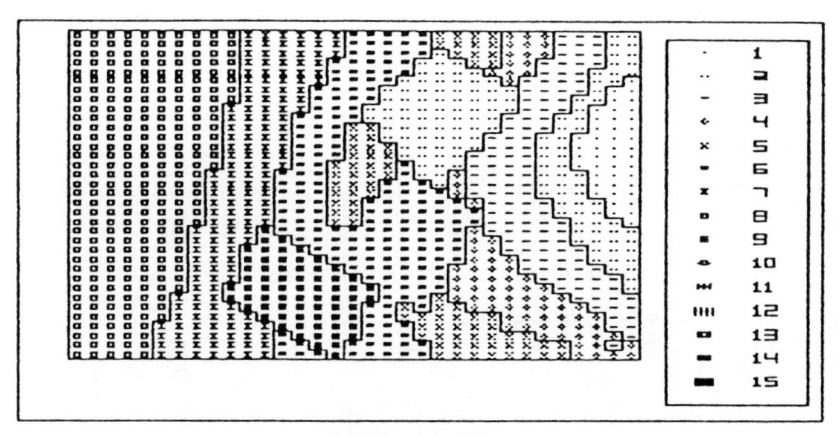

Formation Map

#	Key
1	82 feet
2	83 feet
3	84 feet
4	85 feet
5	86 feet
6	87 feet
7	88 feet

Figure 12-2. Suggested alterations of the natural terrain.

160 CARTOGRAPHY AND SITE ANALYSIS WITH MICROCOMPUTERS

The modified terrain map was then used in conjunction with the original terrain map to estimate the volume and intensity of earthwork excavation and filling involved in making the proposed alterations. Maps depicting areas subject to excavation and filling, the intensities of cutting, and the intensities of filling are shown in figures 12-3, 12-4, and 12-5, respectively.

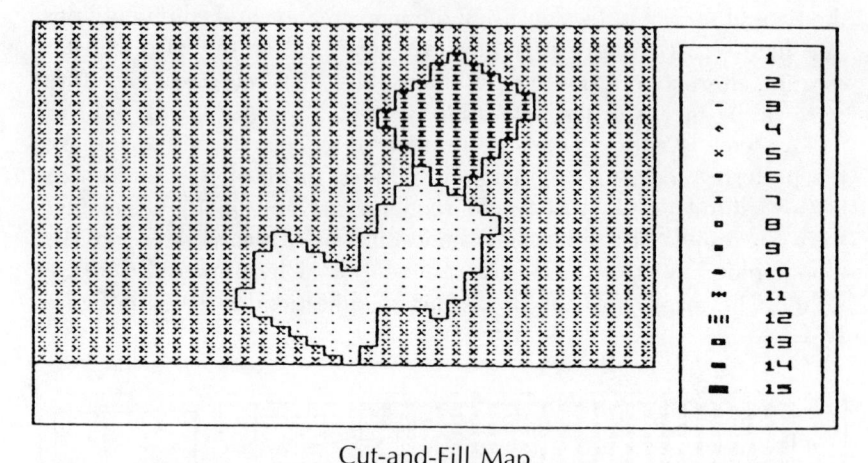

Cut-and-Fill Map

LEGEND

#	Key
5	Neutral
1	Fill (Volume = 230 cubic feet)
7	Cut (Volume = 138 cubic feet)

Figure 12-3. Hard copy of cut-and-fill locations and volumes.

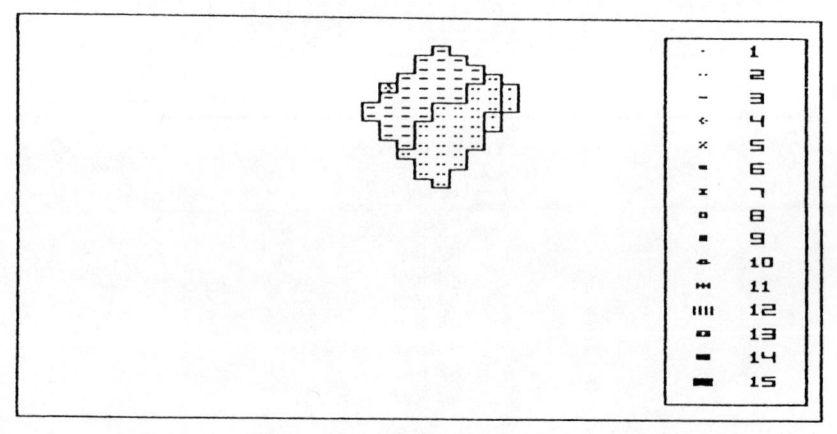

Figure 12-4. Intensity of cutting.

Terrain Modification: Cut-and-Fill Utility   161

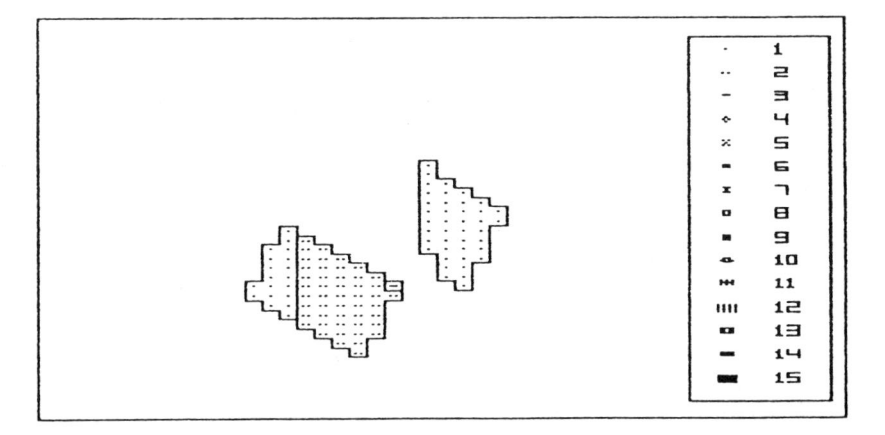

Figure 12-5. Intensity of filling.

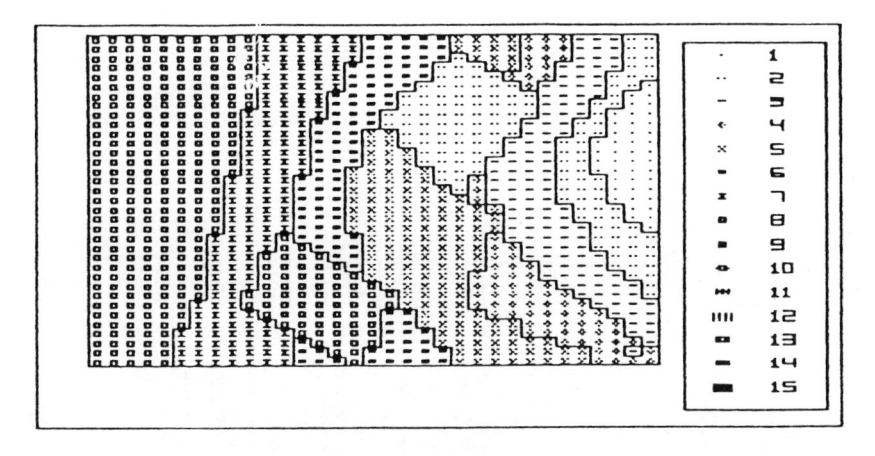

Opt. Formation Map

LEGEND	
#	Key
1	82 feet
2	83 feet
3	84 feet
4	85 feet
5	86 feet
6	87 feet
7	88 feet
8	89 feet

Figure 12-6. Hard copy of the optimum balance of cutting and filling.

162 CARTOGRAPHY AND SITE ANALYSIS WITH MICROCOMPUTERS

It became obvious that the recommended modifications to the site satisfied most design expectations regarding the securing of ideal building sites. These modifications did not, however, attempt to balance cutting and filling as a subsidiary objective of site preparation. This is evident from the fact that the volume of fill exceeded the volume of cut by 92 cubic feet. The recommended modifications thus implied the need for importing 92 cubic feet of earth to the site from elsewhere.

In order to improve the economic viability of the terrain modification process, the optimization option of the cut-and-fill utility was harnessed to balance the volumes of cut and fill. Two constraints for the site were first established: no location (cell) was to have an elevation exceeding 89 feet; and no location (cell) was to have an elevation below 83 feet.

The optimization procedure culminated in the generation of maps showing the optimum (balanced cut-and-fill) terrain modification (see figure 12-6), the volumes of earthwork excavation and filling (see figure 12-7), and the intensities of earthwork excavation and filling (see figures 12-8 and 12-9).

The optimum terrain modification evidently necessitated an export of only 18 cubic feet of earth, thereby rendering the earthwork operation more economical than in the previous alternative.

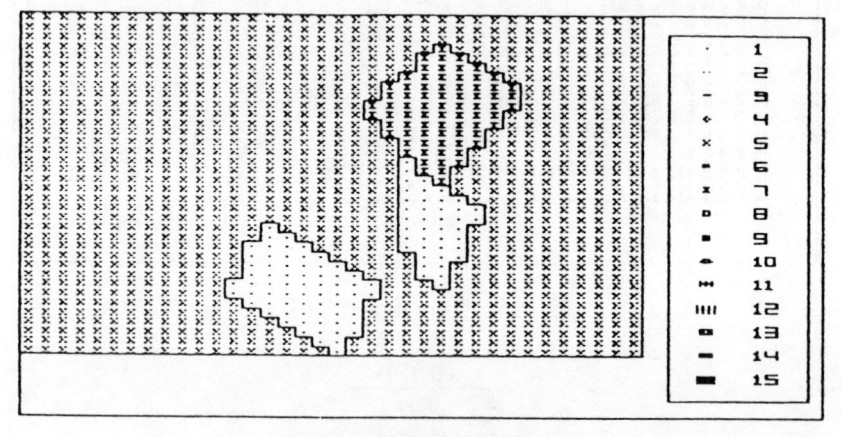

Opt. Cut-and-Fill Map

#	LEGEND Key
1	Fill (Volume = 120 cubic feet)
5	Neutral
7	Cut (Volume = 138 cubic feet)

Figure 12-7. Optimum volume of cutting and filling.

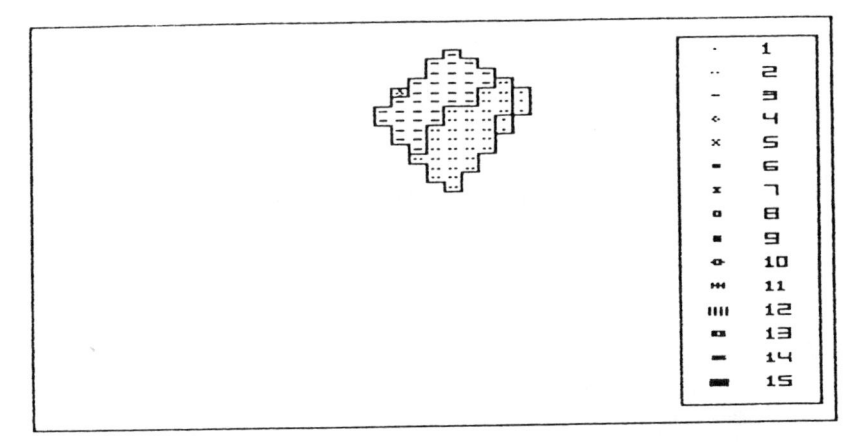

Figure 12-8. Optimum intensity of cutting.

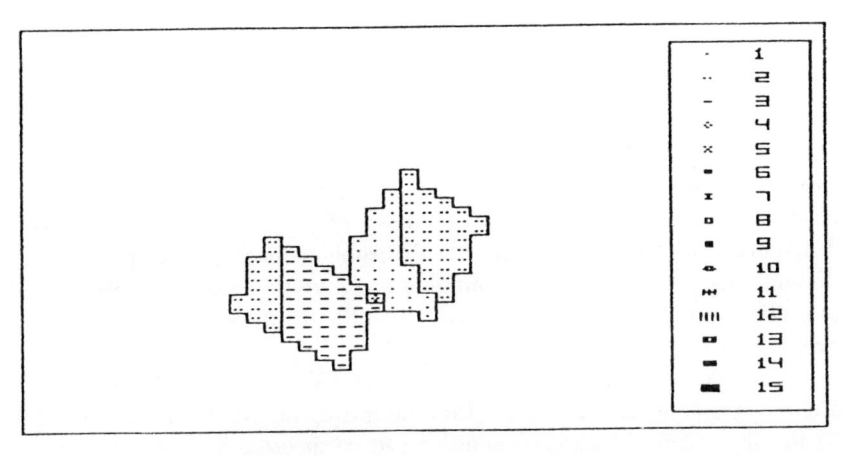

Figure 12-9. Optimum intensity of filling.

A final cautionary note: optimization may not be a critical factor in site preparation for building sites, but it could be crucial in site preparation for large projects such as reservoir construction, park development, and land reclamation. The optimization utility must therefore be used with considerable judgment and discretion.

THIRTEEN

Some Concluding Thoughts

Computers are playing an increasingly valuable role in enhancing the productivity, effectiveness, and efficiency of physical planners, urban designers, and landscape architects in problem-solving. The advent of the microcomputer has contributed greatly to the gradual erosion of inhibitions and the progressive adoption of the technology in these three fields.

An area especially relevant to planners, designers, and landscape architects is computer graphics. Computer graphics has made an impact on the above three disciplines in at least three specialized areas: computer cartography and site analysis; computer-aided design and computer-aided drafting; and schematic and statistical data representation. Hardware and software continue to develop at a staggering pace, suggesting that technological power and sophistication are now available to professionals at a very inexpensive price.

Several utilities intended for computer cartography and site analysis applications are developed and demonstrated in this book. The list of such utilities is far from comprehensive, and many more remain to be developed. Thus, this book only makes a start, leaving the list of utilities open-ended. Readers may design and develop a number of extensions and adaptations to the utilities discussed in this book, and may add a number of new ones.

Some Concluding Thoughts 165

Although they are the focus of this book, computer cartography and site analysis applications constitute just one of three varieties of graphics applications that are relevant to an audience of physical planners, urban designers, and landscape architects. Computer-aided drafting and computer-aided design (CAD) applications constitute another variety of graphics applications—one that is currently experiencing phenomenal growth and advancement.

CAD enables a user to undertake both two- and three-dimensional drafting and design. Two-dimensional drafting with CAD equipment and software considerably enhances the user's speed, productivity, and precision in creating production and shop drawings.

Three-dimensional drafting and design allow the user to visualize, design, and draft in three dimensions with relative flexibility and creative freedom. Geometric points, lines, and planes may be created and transformed into images of three-dimensional objects and surfaces rapidly and accurately. These images may be saved in a symbol library for future retrieval and use in creating a layout, designing a floor plan, or rendering an elevation.

Three-dimensional images may be viewed in two- and three-point perspectives, and plans and layouts may be subjected to critical visual scrutiny and analysis. The images lend themselves to manipulation and transformations whereby the products of layout planning and design are modified, adjusted, or dramatically changed in a dynamic, interactive graphics environment. The time taken to complete these changes on computer is a bare fraction of that of the corresponding manual process.

Computer images generated on a CAD system also lend themselves to coupling with "bill-of-materials" applications. A bill of materials application enables a user to link a graphics image of an object with considerable descriptive information pertaining to the object. This information may then be organized and compiled into bills of materials for cost estimation and facilities management.

Solids-modeling—another graphics application that is rapidly gaining prominence—deals with solid objects having opaque surfaces. Images of objects such as cubes, cones, cylinders, spheres, and pyramids are added, subtracted, and manipulated to create new images of composite objects.

Computer graphics is not the only area in which the capabilities of a microcomputer may be harnessed by a planner or designer. The microcomputer is a general-purpose computing device; the choice of software and its capabilities enable the user to manage special applications in response to a variety of specific needs. For instance, word processing software may be used professionally for reports, business correspondence, contract documentation, specifications writing, and the like. Present-day word processors are equipped with spelling checkers and mailing list capabilities that have completely transformed the manner in which professional office correspondence is conducted.

Spreadsheet software can serve the planning/design office in a number of ways. Various spreadsheet templates have been designed and marketed for specialized computational applications. Planners use spreadsheets for forecasting, statistical data processing, simulation models, and other applications. Designers and landscape architects use spreadsheets for performing design calculations, materials and cost estimations, and so on. Current-generation spreadsheet software is integrated with data base management and business graphics capabilities, considerably enhancing the versatility of their application.

Data base management is a common requirement of any professional office. Data base management software enables a user to file voluminous data and to retrieve information in report format, based on specific criteria. Typical applications of data base management software are inventories, lists of materials, client and project data, and bibliographies of reference literature.

Project management software enables users to perform critical path analyses (CPM) and program review and evaluations (PERT). Project scheduling and manpower and resource allocation are additional tasks relevant to planning, design, and landscape architecture that may be undertaken with project management software.

Finally, networking, communications, and electronic mail constitute three extremely powerful microcomputer capabilities. Networking and communication are made possible through hardware devices such as modems, telephone lines, and direct lines between computers. Networks of computers may be formed within a professional office or among several locations, nationwide. Communications software enables one computer to exchange information, share data, transfer files, and transmit and retrieve messages with another computer.

Computer hardware has entered a period of tremendous flux, and technological breakthroughs occur with remarkable frequency. The earliest microcomputers were equipped with 8-bit microprocessors; the 16-bit microprocessor has since become the industry standard. The 32-bit or 64-bit desktop microcomputer is already in development and appears to be realizable in the very near future. The increasing power of microprocessors, microcomputers' rapidly expanding memory capabilities, and the moderateness of their cost make microcomputers versatile, powerful, and affordable tools.

The competitive edge in professions affected by this new technology appears to hinge on the availability and judicious application of good, user-friendly software. Such software should enable planners and designers to concentrate on the problem-solving process without having to contend with complex data structures, integrated programming, and file management. Developments in such fields as artificial intelligence and expert systems, laser disk and video disk technology, and image enhancement and analysis may eventually filter into the planning and design fields after suitable adaptation and testing. Practitioners in these fields must therefore make an early start in getting acquainted with current-generation hardware and software—not only to establish a competitive edge today, but to ensure competitiveness in the future, as new technological advances occur.

The principles, techniques, and methodologies explained and illustrated in this book offer a reasonable start in the use of microcomputer graphics for thematic mapping and site analysis. The fundamental logic of these principles, techniques, and methodologies is not likely to undergo dramatic changes, although enhancements and improvements that may influence the conduct of specific processes and procedures are quite possible.

A good start at the formative stage, together with hands-on experience, will contribute toward considerable progress and a competitive edge as the sophistication of microcomputer technology enters a new era of development.

APPENDIX A

Accessing the Graphics Tablet

This appendix is furnished to enable the user to secure a basic understanding of the dynamics of graphics tablets.

THE ANATOMY OF A GRAPHICS TABLET

A graphics tablet (in this instance, an Apple Graphics Tablet) is equipped with an interface card containing 2K bytes of ROM. This ROM includes all the subroutines that read and interpret the signals from the graphics tablet; the subroutines can be accessed from a BASIC program.

One principal purpose of these subroutines is to read the position of the stylus in relation to the surface of the graphics tablet. The subroutines also perform several other functions:

1. They provide a means of selecting the desired screen mode from ten alternative choices, ranging from text mode to high-resolution graphics mode.
2. They enable a user to select and assign a scaling divisor—ranging in value from 1 to 32,767—to regulate and synchronize stylus and cursor movement.
3. They read the signals from the stylus, in order to detect: whether the stylus has moved from its previous location by a readable

distance; whether the stylus is pressed down on the tablet or not; whether the stylus is active in the stream mode or not; and whether a key has been pressed on the keyboard or not.
4. They regulate the display of the cursor on the screen, subject to the syntax of assigned scaling and offset information.
5. They help calibrate the position of the origin on the graphics tablet. Horizontal and vertical offset information about the stylus location (relative to that of the anticipated origin on the tablet) is supplied to the computer by the user during the calibration process. The numerical values of offsets are integer numbers lying within the range −32,767 to 32,767.
6. They make possible the suppression of output on the screen.

COMMUNICATING WITH THE GRAPHICS TABLET

Communication with the graphics tablet is accomplished by accessing the subroutines resident on ROM. A BASIC program using the following two commands may be used to permit this communication:

```
PR# s

IN# s
```

In the above commands, *s* represents the slot number into which the graphics tablet interface card is inserted.

The *PR# s* command directs all control commands to the graphics tablet. The *IN# s* command reads all succeeding inputs from the tablet.

The variable *s* is assigned a value of zero only under the following circumstances. If control commands are to be transmitted to the screen instead of to the graphics tablet, *PR#0* must be used. Similarly, if communication from the graphics tablet is to be terminated and if succeeding inputs are to be derived from the keyboard, *IN#0* must be used.

SECURING CONTROL OF THE GRAPHICS TABLET

Control of the graphics tablet may be achieved by executing a BASIC program containing a *PR# s* command and a PRINT string of tablet control commands.

There are seventeen tablet control commands, all of which are described completely in the documentation accompanying the graphics tablet. Some of these commands use the integer argument *i*.

GR i or **G i** sets the display mode of the monitor screen to low-resolution graphics. Four lines of text are permitted below the low-resolution graphics display. The argument *i* assigns the page number for text and graphics and may have a value of either 1 or 2.

SCALE i or **S i** sets the tablet scaling divisor to *i*. All coordinates generated by the graphics tablet are divided by *i* before delivery to the program for processing. The argument *i* may have a value from 1 to 32,767. Negative values are converted to positive values, and zero is invalid.

XOFF i or **X i** sets the horizontal offset (X axis) of the tablet to *i*. The offset value of *i* may range from −32,767 to 32,767.

YOFF i or **Y i** sets the vertical offset (Y axis) of the tablet to *i*. The offset value may range from −32,767 to 32,767.

R represents an option that may be used in conjunction with the scaling divisor SCALE. All coordinates generated by the graphics tablet are divided by the scaling divisor SCALE before they are delivered to the BASIC program for processing. Offset values are added only after this operation is performed.

BEFORE or **B** is an option that may be used in conjunction with the R command. When it is selected, the offset values are added before the scaling operation (mentioned above) is performed.

NOPRINT or **N** disables transmission of output to the screen. *NOPRINT* may be deactivated by using the *PR#0* command.

The use of control commands in a BASIC program are subject to the following parameters: commas may be used as delimiters between commands, provided that a command string does not begin or end with a comma; and only the first alphabetic character of a command string is used, while all spaces and subsequent alphanumeric characters are ignored.

PROGRAMMING WITH CONTROL COMMANDS

Control commands may be programmed in BASIC using a particular syntax as follows. If the tablet interface card is located in slot number *SL*, if the tablet is to be set to the low-resolution graphics mode (with four lines of text below it), if a scaling divisor *SF* is to be used, if *XF* and *YF* are assigned as offset values, if the offset values are to be added before the scaling operation, and if screen display is disabled, then the following is the programming syntax:

```
PR#  SL:  PRINT"GR 1,  SCALE";  SF;",
XOFF=";XF;",  YOFF=";YF;",  R, B, N"
```

It is important not to add a semicolon (;) or comma (,) at the end of the above *PRINT* string. A semicolon or comma will suppress the *RETURN* character, which is essential for executing any command. In the absence of the *RETURN* character, the command cannot be executed by the tablet.

The above syntax may be abbreviated by eliminating blank spaces and using only the first letters of each control command name, as follows:

```
PR#SL: PRINT"G1,S";SF;",X";XF;",Y";YF;",R,B,N"
```

The following listing is an example of how the above control command string would appear in a BASIC program:

```
10 D$= CHR$(4)
20 PRINT D$;"PR#";SL: PRINT "G1,S";SF;",X";XF;
   ",Y";YF;",R,B,N"
30 PRINT D$;"PR#0"
```

The *PR#0* command is used after the control command string in order to redirect output to the screen.

INTERPRETING INPUT

The preceding commands orient the graphics tablet for the type of numerical inputs that may be derived from the tablet. The BASIC commands *IN#* and *INPUT* may now be used to derive the coordinates and status information of the stylus from the tablet.

The tablet transmits coordinate and status information about the stylus, using the following format:

```
X-position, Y-position, sign status
```

In this format, X-position and Y-position are coordinates having integer values between $-9,999$ and $9,999$ (inclusive), and sign status represents the status of the stylus and the keyboard.

The X-position and Y-position coordinates indicate the position of the stylus on the tablet. If the *R* command is in effect, the offset is added and the result is divided by the scale to derive coordinate information about the stylus. The user must make certain that the X-position and Y-position offsets returned by the tablet and the scale fall within the prescribed range of integer values; if this range is exceeded, an error occurs.

CALIBRATION OF THE TEMPLATE

An important prerequisite of using the graphics tablet is to calibrate the template. Calibration is the process by which the active work space of the graphics tablet corresponds accurately and precisely to the area shown on the screen.

The procedure for calibration is explained in detail in the documentation accompanying the graphics tablet; some adjustments in the calibration procedure need to be made to customize it to the user-designed template. The user must provide coordinate information about the top left-hand corner and the bottom right-hand corner of the user-designed template during the calibration process. This information, together with the slot number (assigned to the graphics tablet interface card), is saved for future use in determining XOFF, YOFF, and SCALE.

Registered Trademarks

The following items bearing registered trademarks, as shown below, have been cited in this book.

Lisa, Apple Business Graphics, Apple II Plus and Apple IIE are trademarks of Apple Computer, Inc.

CDC is the trademark of Control Data Corporation.

IBM is the trademark of IBM Corporation.

PDP-II and VAX are trademarks of Digital Equipment Corporation.

Radio Shack Personal Computer is the trademark of Tandy Corporation.

CP/M, CP/M 2.2, CP/M+, CP/M 86, and MP/M are trademarks of Digital Research.

MSDOS 1.2 and MSDOS 2.0 are trademarks of Microsoft Corporation.

OASIS and OASIS-16 are trademarks of Phase One System, Inc.

SYMAP is the trademark of Laboratory for Computer Graphics and Spatial Analysis at Harvard University.

IMGRID was developed by David F. Sinton at the Harvard Graduate School of Design.

IMPAC was developed by Egbert Scientific Software and marketed by Spectral Data Corporation.

GRIDAPPLE was designed by Environmental Systems Research Institute and distributed by IRIS International, Inc.

PC Atlas is the trademark of Strategic Locations Planning.
STATMAP is the trademark of Ganesa Group International.
PC MAP is the trademark of Geo-Software.
RIPS is the trademark of EROS DATA CENTER.
APPLEPIPS is the trademark of Telesys Group.
ERDAS 400 is the trademark of ERDAS, Inc.
HOTLIPS is the trademark of Nebraska Remote Sensing Center.
APPLE/APP is the trademark of the Department of Geography, University of Georgia.
LMS is the trademark of Measuronics Corporation.
CADAM is a registered trademark of CADAM, Inc.
CALCOMP IGS-400 and IGS-500 are the trademarks of California Computer Products, Inc.
McAuto General Drafting System is the trademark of McDonnell Douglas.
Intergraph Architectural and Engineering Application System is the trademark of Intergraph Corporation.
Designer IV, V, and M are trademarks of Computer Vision Corporation.
AutoCAD is the trademark of AutoDesk, Inc.
AE/CADD was created by Archsoft Corporation.
CADKEY is the trademark of Micro Control Systems.
CADPLAN was designed by Personal CAD Systems.
Microcad is the trademark of Computer Aided Design of San Francisco.
CAD-1 and CAD-2 are trademarks of Chessel-Robocom Corporation.
CASCADE I and II are trademarks of Cascade Graphics Development.
CADAPPLE is the trademark of T&W Systems, Inc.
SPACE GRAPHICS/SPACE TABLET is the trademark of Micro Control Systems.
6502 is the trademark of SYNERTEK.
MC 68000 is the trademark of Motorola, Inc.
UNIX is the trademark of Bell Laboratories.
UCSD-P is the trademark of Softech Microsystems.
SAS/GRAPH is the Trademark of SAS Institute, Inc.
SPSS Graphics, SPSS, and SPSS X are the trademarks of SPSS, Inc.
TEKTRONIX is the trademark of TEKTRONIX, Inc.
Hewlett-Packard is the trademark of Hewlett-Packard, Inc.
Executive Presentation Kit is the trademark of Koala Technologies Corporation.
Business Graphics Systems is the trademark of Peachtree Software, Inc.
PFS: GRAPH is the trademark of Software Publishing.

Registered Trademarks 175

LOTUS Graphics is the trademark of Lotus.
Microsoft Chart is the trademark of Microsoft Corporation.
dGRAPH III was designed by Fox and Geller.
EPSON MX5OFT with GRAFTRAX is the trademark of EPSON America, Inc.

This list is subject to correction if any errors are brought to the attention of the authors and/or the publishers.

Bibliography

Anjomani, Ardeshir. "The Overlaying Map Technique: Problems and Suggested Solutions." *Journal of Planning Education and Research* 4(2) (December 1984): 111–19.

Apple Computer, Inc. *Applesoft: BASIC Programming Reference Manual.* Cupertino, Ca.: Apple Computer, 1981.

———. *Apple II DOS Manual: Disk Operating System.* Cupertino, Ca.: Apple Computer, 1981.

———. *Apple II Reference Manual: A Reference Manual for the Apple II and the Apple II Plus Personal Computers.* Cupertino, Ca.: Apple Computer, 1981.

———. *Graphics Tablet: Operation and Reference Manual.* Cupertino, Ca.: Apple Computer, 1981.

Architectural Record (October 1984): 49–80.

Buffa, E. S.; Armour, G. C.; and Vollman, T. E. "Allocating Facilities with CRAFT." *Harvard Business Review* 42 (March–April 1964).

Cerny, James W. "Use of SYMAP Computer Mapping Program." *Journal of Geography* 71(3) (March 1972): 167–74.

Demel, John T., and Miller, Michael. *Introduction to Computer Graphics.* Belmont, Ca.: Wadsworth, 1984.

Dickinson, John. "Business Graphics: Giving Power to Presentations." *PC: The Independent Guide to IBM Personal Computers* 4(12) (June 1985): 110–237.

Bibliography

Egbert, Dwight D. "Low Cost Image Analysis: A Microcomputer Based System." *Computer Graphics World* 5(2) (February 1982): 58–59.

Giloi, Wolfgang K. *Interactive Computer Graphics: Data Structure, Algorithms, Languages.* Englewood Cliffs, N.J.: Prentice-Hall, 1978.

Harrington, Steven. *Computer Graphics—A Programming Approach.* New York: McGraw-Hill, 1983.

Horn, Carin E., and Poirot, James L. *Computer Literacy: Problem Solving with Computers.* Austin, Tx.: Sterling Swift, 1981.

MacDougall, E. Bruce. *Microcomputers in Landscape Architecture.* New York: Elsevier Scientific Publishing, 1983.

McHarg, Ian L. *Design with Nature.* Garden City, N.Y.: Doubleday, 1971.

Mitchell, William. *Computer Aided Architectural Design.* New York: Petrocelli/Charter, 1977.

Monmonier, Mark S. *Computer Assisted Cartography: Principles and Prospects.* Englewood Cliffs, N.J.: Prentice-Hall, 1982.

Myers, Roy E. *Microcomputer Graphics.* Reading, Ma.: Addison-Wesley, 1982.

Omaha–Council Bluffs Metropolitan Area Planning Agency. "Environmental Resource Analysis Program: A Resource Information Base." Omaha: Omaha–Council Bluffs MAPA, 1979.

Park, Chan S. *Interactive Microcomputer Graphics.* Reading, Ma.: Addison-Wesley, 1985.

Poirot, James L. *Microcomputer Systems and Apple Basic.* Austin, Tx.: Sterling Swift, 1980.

School of Environmental Development. "Reference Manual for Synagraphic Computer Mapping: SYMAP." Version V, Draft #2. Lincoln, Nebr.: University of Nebraska.

Scripter, Morton W. "Choropleth Maps on Small Digital Computers." *Proceedings of the Association of American Geographers* 1 (1969): 133–36.

Sinton, David F. *An Introduction to I.M.G.R.I.D.: An Information Manipulation System for Grid Cell Data Structures.* Department of Landscape Architecture, Harvard University (June 1976).

Smith, Dennis R.; Smith, Gary R.; and Coffey, Catherine M. "Formalizing Geographic Information Processing Methods: Land Use Suitability Mapping." *Computer Graphics World* 5(9) (September 1982): 34–44.

Steck, Richard. *Apple to IBM PC Conversion Guide.* Glenview, Il.: Scott, Foresman, 1985.

Teague, Lavette C. "Network Models of Configurations of Rectangular Parallelopipeds." In *Emerging Methods in Environmental Design and Planning*, Gary T. Moore, ed. Cambridge, Ma.: MIT Press, 1970.

Welch, Roy A.; Jordan, Thomas R.; and Usery, E. Lynn. "Microcomputers in the Mapping Sciences." *Computer Graphics World* 6(2) (February 1983): 34–36.

Index

active work space, 45–46
AE/CADD, 2, 34
Apple, 35–36
   Graphics Tablet, 41–42
   Macintosh, 16
   II Plus, 16, 41
Apple/APP, 33
APPLEPIPS, 32
Archsoft Corporation, 34
AutoCAD, 2, 34
AutoDesk, Inc., 34
auxiliary storage devices. See
   Mass storage devices

base unit, 12
BASIC, 20, 41
Bell Laboratories, 19
bit, 16
business graphics, 28, 164–65
   software for, 3, 36–38
Business Graphics System, 3, 37
byte, 16

C, 20
CADAM, 33
CADAM, Inc., 33
CADAPPLE, 35–36
CADKEY, 2, 34–35
CAD-1, 35
CADPLAN, 3, 35
CAD-2, 35
Calcomp, 2
   IGS-400, 33
   IGS-500, 33
   System 25, 33
California Computer Products, 33
camera systems, 16
cartography. See Computer cartography; Thematic mapping
Cascade Graphics Development, 35
CASCADE-I, 35
CASCADE-II, 35
catalog utility, 89–90
cell protection utility, 68–69

cell splitting, 43–44
cell utility, 57–58
Chessel-Robocom Corporation, 35
clear screen utility, 90–91
clock cycle, 16, 17
CMAP, 2, 29
COBOL, 20
color selection program, 55–57
compatibility, 19
compilers, 20
composite video monitors, 14, 22, 41
Computer Aided Design, 35
Computer-aided drafting/design (CAD), 27–28, 165
   benefits of, 2
   software for, 2–3, 33–36
computer cartography, 26–27. *See also* Thematic mapping
   benefits of, 2
   dynamics of main routine program for, 51–54
   graphics tablet template for, 45–46
   main routine program for, 46–50
   representing spatial data with, 42–44
   software for, 2, 28–33
   useful utilities for, 40–41
computer graphics. *See also* Business graphics; Computer-aided drafting/design; Computer cartography
   advantages of, 25
   applications of, 26–28
   defined, 25
   imagery classifications for, 25–26
computers. *See also* Microcomputers
   for business graphics, 3, 36

for CAD, 2, 33–34
for cartography, 2, 28–30
in design, 1
early antipathy toward, 1–2, 7–8
graphics applications of, 9
in planning, 1
Computer Vision Corporation, 2, 33
CP/M, 19
cut-and-fill utility
   algorithm for, 151
   case study application of, 159–63
   cautions for using, 150
   objective of, 149
   optimization in, 150, 151–52, 163
   procedure for, 150–51
   subroutine for, 153–58

data, 20
Data Management, 30
delete utility, 90
Designer
   IV, 33
   V, 33
   M, 33
Design with Nature (McHarg), 100
dGraph III, 38
Digital Research, 19
digitizers, 13, 14, 23, 44
digitizing tablets. *See* Digitizers
directory, 18
disk drives, 12
disks, magnetic, 18, 22
dot matrix printer, 15, 22
drawing management utilities
   applications of, 91–92
   function of, 73

HARD COPY program for, 77–86
  legend option, 74
  subroutine for catalog, 89–90
  subroutine for clear screen, 90–91
  subroutine for delete, 90
  subroutine for legend, 86–89
  subroutine for load, 75–76
  subroutine for print, 77
  subroutine for save, 74–75
  types of, 73–74
draw utility, 59

Earth Resources Data Analysis, 32
Egbert Scientific Software, 32
electronic mail, 23
electrostatic plotters, 15
Environmental Systems Research Institute (ESRI), 30
EPSON MX80FT, 42
erasable programmable read only memory (EPROM), 17
EROS 400, 32
EROS Data Center, 32
evaluation utility
  algorithm for, 123
  case study application of, 125–40
  purpose of, 122–23
  subroutine for, 124–25
Executive Presentation Kit, 3, 37

Fisher, Howard T., 29
flatbed plotters, 15, 23
flat panel displays, 14
floppy disks, 18, 22
FORTH, 20
FORTRAN, 20
Fox and Geller, 38

Ganesa Group International, 31
Geo-Software, 31
Giloi, Wolfgang, 26
graphics tablets. *See* Digitizers
graphics template, 45–46
  main routine for, 46–50
GRIDAPPLE, 2, 30–31
grid mapping, 42–43

HARD COPY program, 77–86
hard disks, 18, 22
Harvard Graduate School of Design, 30
Harvard Laboratory for Computer Graphics and Spatial Analysis, 29
high-level languages, 20
HOTLIPS, 32

IBM
  PC AT, 16, 34
  PC, 19, 34–35
  PC/XT, 19, 34
  360, 30
  370, 30
icons, 14
image analysis, 31–33
IMGRID, 2, 30, 42
IMPAC, 32
ink jet printers, 15
input devices, 13, 14
input/output bus, 16
Intergraph Architectural and Engineering Application System, 33
Intergraph Corporation, 2, 33
internal system bus, 16
I/O device, 13–14

joysticks, 13

keyboards, 14

key pad, numerical, 23
kilobyte, 16
Koala Technologies Corporation, 37

languages, 19–20
laser printers, 15, 22–23
legend utility, 86–89
letter-quality printers, 15, 22
light pens, 13, 14, 23
line utility, 60–61
LISP, 20
LMS System, 33
load utility, 75–76
Lotus Development Corporation, 37
LOTUS 1-2-3, 3, 37
low-level languages, 20

McAuto General Drafting Systems, 33
McDonnell Douglas, 33
McHarg, Ian, 100
machine code, 19–20
mapmaking. See Computer cartography
mass storage devices, 13, 17–18
Measuronics Corporation, 33
megabyte, 17
memory. See Primary memory
menu command area, 45–46, 51–53
MICROCAD, 2, 35
microcomputers
  acceptance of, 2, 8
  for business graphics, 37–38
  for CAD, 2–3, 34–36
  for cartography, 2, 30–31
  dedicated, 12, 21
  defined, 11
  elements of system configuration in, 12–13
  general-purpose, 12
  organization of, 13–18
  power of, 13
  selecting configuration of, 20–23, 24
Micro Control Systems, 34, 35
microprocessors, 13, 16–17
Microsoft CHART, 3, 37–38
Microsoft Corporation, 19, 37
modems, 13, 23
monitors
  display screen of, 14–15, 22
  resolution of, 15, 22
  types of, 14
monochrome monitors, 14, 22
mouses, 13, 14, 23
MS DOS, 19
multiple-map overlay utility, 100
  assigning attribute weights in, 101–2, 116–20
  case study applications of, 112–21
  features of, 101
  off-line procedures for, 101–2
  on-line procedures for, 103–10
  subroutine for, 110–12
multiple-pen plotters, 15–16

Nebraska Remote Sensing Center, 32

OMNIBUS, 30
opaque overlay utility, 96–98, 99, 100
operating system, 19, 21
output device, 13, 14–16

PASCAL, 20
PC Atlas, 2, 31
PC DOS, 19
PC MAP, 2, 31
Peachtree Software, 37

Index

pen plotters, 15
Personal CAD Systems, 35
PFS: Graph, 3, 37
photogrammetry, 31–33
plasma monitors, 14
plotters, 12, 15–16, 23
polygon mapping, 42–43, 44
polygon utility, 62–65
primary memory, 13, 17, 22
printers, 12, 15, 22–23
print utility, 77–86
programmable read only memory (PROM), 17
PROLOG, 20

random access memory (RAM), 17
read only memory (ROM), 17
recode utility, 70–72
remote sensing, 31–32
resolution, 15, 22, 43–44
RGB monitors, 14, 22
RIPS, 32

SAS/GRAPH, 3, 36
SAS Institute, 36
save utility, 74–75
Scripter, Morton, 29
simple scan utility, 93–95, 98–99
single-pen plotters, 15–16
Sinton, David F., 30
Softech Microsystems, 19
software
 applications of, 18–19, 20
 for business graphics, 3, 20
 for CAD, 2–3, 20, 33–36
 and choice of microcomputer, 21
 for computer cartography, 2, 28–33
 defined, 18
 for image processing, 32–33
 system program, 18–20
 user-friendliness of, 44
Software Publishing, 37
SPACE GRAPHICS/SPACE TABLET, 35, 36
Spectral Data Corporation, 32
SPECDATA, 32
SPSS Graphics, 3, 36–37
SPSS Inc., 36
STATMAP, 2, 31
storage devices. *See* Mass Storage devices
Strategic Locations Planning, 31
subroutine utilities
 catalog, 89–90
 cell, 57–58
 cell protection, 68–69
 clear screen, 90–91
 color selection, 55–57
 cut-and-fill, 153–58
 delete, 90
 draw, 59
 evaluation, 124–25
 legend, 86–89
 line, 60–61
 load, 74–76
 multiple-map overlay, 110–12
 opaque overlay, 96–97
 polygon, 62–65
 print, 77
 recode, 70–72
 save, 74–75
 scan, 93–95
 visibility analysis, 143–45
SYMAP, 2, 29, 30, 42

T & W Systems, Inc., 35
tape drives, 13
tapes, magnetic, 18
telecommunications, 23, 168
Telesys Group, 32

television sets, 14
thematic mapping
 assigning colors for, 67
 defined, 4
 digitizing source map for, 67–68
 editing, 68–72
 parameters for, 66–67
 subroutine for cell utility, 57–58
 subroutine for draw utility, 59
 subroutine for line utility, 60–61
 subroutine for polygon utility, 62–65
 subroutine for selecting colors, 55–57
thermal printer, 15
time-sharing systems, 19, 21
TOMLIN, 30

UCSP-P, 19
United States Census Bureau, 29
University of Georgia, 33
UNIX, 19
user memory. *See* Primary memory
user preparedness, 9–10

vertical plotters, 15, 23
video cameras, 13
visibility analysis utility
 application of, 145–47
 principle behind, 141–42
 procedure for using, 142–43
 purpose of, 141
 subroutine for, 143–145

word size, 16, 17

XENIX, 19